Contents

Foreword

This publication is one of the outcomes of a workshop on Rice Tungro Disease held in Alor Setar, Malaysia, 11–14 November 1993, and organized by a committee headed by Dr Supaad Mohd. Amin of the Malaysian Agricultural Research and Development Institute (MARDI). The organizing committee was aided by the Natural Resources Institute (NRI), UK, and the International Rice Research Institute (IRRI) who also provided financial support. Additional sponsorship was given by the Food and Agriculture Organization of the United Nations, the Swiss Development Co-operation and the United Nations Development Programme.

Fifty delegates from India, Indonesia, Japan, Malaysia, Nepal, Philippines, Thailand and United Kingdom participated in the workshop. The objectives of the workshop were to review the current status of rice tungro virus disease in South and South-East Asia and to discuss ways of formulating appropriate tungro management strategies in the light of the new research information available. Much of this information has resulted from a collaborative project on the vector ecology, epidemiology and management of rice tungro disease begun by NRI and IRRI in October 1990. Important findings have also been made in recent studies carried out by the Indonesian Department of Agriculture in collaboration with the Japan International Co-operation Agency on the epidemiology and forecasting of tungro.

During the workshop formal presentations were made by country representatives and by scientists on specific areas of tungro research. Working groups critically examined current tungro management strategies and developed recommendations for further research and strategy evaluation. These recommendations mark a significant advance in the establishment of collaborative efforts between national and international institutes to develop tungro management strategies based on sound epidemiological principles and in accordance with integrated pest management practices. Many of the recommendations detailed in the action plans agreed at the workshop have already been taken up in new collaborative research programmes.

We would like to thank the members of the organizing committee for their unstinting efforts in arranging the meeting. Our sincere appreciation is expressed to our hosts at the Muda Development Authority and to the facilitators who ensured the smooth running of the sessions. Drs K.L. Heong and P.S. Teng of IRRI were instrumental in initiating the workshop and played important co-ordinating roles. Thanks are also due to Ms Ellen Genil for secretarial and logistical support at IRRI.

T.C.B. Chancellor
J.M. Thresh

Preface

For almost three decades, rice tungro virus disease has been widely recognized as a serious constraint to rice production in intensively cultivated, irrigated areas in a number of countries in South and South-East Asia. Although the frequency and intensity of tungro outbreaks have declined since the peak in the late 1960s and the early 1970s, occasional major epidemics still occur and the disease is endemic in certain areas. The potential of tungro to cause severe yield loss and the lack of effective control measures available to rice farmers account for the continuing high profile of the disease. Nevertheless, there are many irrigated rice areas where tungro is unimportant or entirely absent. In such areas, insecticide-based strategies to control the leafhopper vectors are often recommended although they are difficult to justify on current evidence. Moreover, insecticide usage commonly poses a risk to human health and threatens to undermine the success of integrated pest management programmes.

In this workshop, reports from representatives of five major rice producing countries in South and South-East Asia, where tungro outbreaks have occurred in the past, indicated that the disease is currently viewed as a serious problem only in India and the Philippines. However, many participants emphasized the sporadic unpredictable nature of the disease and considered that tungro continues to pose a potential threat to rice production and yield stability in the region as a whole. This was demonstrated in the 1995 wet season in Central Java, Indonesia, when a major tungro outbreak occurred causing extensive damage to rice crops. The widespread cultivation of a highly susceptible variety, Cisadane, is thought to have been a major contributory factor to the outbreak in Central Java. This highlights the difficulties in implementing certain tungro control measures such as the cultivation of resistant varieties. Lack of adoption may be due to infrastructure constraints or to socio-cultural factors such as the varietal preferences of farmers. This was a recurrent theme in the reports of the workshop working groups, and potential constraints to adoption of actual or potential tungro control measures are summarized in the earlier report.

Resistance breeding and the problems of finding durable resistance to tungro disease are covered here by three papers. The focus of current breeding programmes, such as that at IRRI, on incorporating resistance or tolerance to tungro viruses in germplasm is considered in relation to variation in virus isolates between different areas in the region. The workshop recommended that multilocational testing of selected lines should be conducted, using revised screening techniques under a newly constituted network of collaborators. The need for studies in variation in populations of leafhopper vectors and in tungro viruses was identified. Further recommendations were made for the conduct of detailed research on leafhopper feeding behaviour and on virus resistance mechanisms at the cellular level.

The development of rapid, simplified methods for detecting tungro viruses in rice plants is summarized in one paper. Recent advances in the knowledge of tungro epidemiology from detailed field studies are discussed in two papers. These are complemented by a paper on the use of simulation modelling to investigate tungro dynamics within individual plantings and to assess the potential of insecticides and roguing to reduce disease incidence. Substantial progress has been made in our understanding of tungro epidemiology in the last few years. We believe that this knowledge provides a firm basis for the development and implementation of appropriate tungro management strategies. In two papers in the book, current options for tungro management are outlined and vector control strategies are critically reassessed.

We hope that the workshop and this book will help to summarize and disseminate the developments in recent research on tungro and the current status of the disease. In addition, we hope that they will act as a catalyst for further collaborative research between national and international institutes with a view to enhancing the prospects for effective and sustainable tungro control strategies.

Abbreviations used in this report

AARD	Agency for Agricultural Research and Development
ASEAN	Association of South East Asian Nations
ATI	Agricultural Training Institute, Philippines
BCKV	Bidhan Chandra Krishi Viswavidyalaya, West Bengal, India
BPH	brown planthopper
BPMC	butyl phenyl methyl carbamate
BSA	bovine serum albumin
DAS	days after sowing
DAS-ELISA	double-antibody sandwich-ELISA
DAT	days after transplanting
DFID	Department for International Development
DOA	Department of Agriculture, Malaysia
ELISA	enzyme-linked immunosorbent assay
ETL	economic threshold level
FOA	Farmers' Organization Authority, Malaysia
GLH	green leafhopper
GOI	Government of Indonesia
IAAS	Institute of Agriculture and Animal Sciences, Nepal
IgG	immunoglobulin
IPM	integrated pest management
IRGC	International Rice Germplasm Collection, IRRI
IRRI	International Rice Research Institute, Philippines
IVI	infective vector index
JII	John Innes Institute, UK
KADA	Kemubu Agricultural Development Authority, Malaysia
LGUs	local government units
MADA	Muda Agricultural Development Authority, Malaysia
MARDI	Malaysian Agricultural Research and Development Institute, Malaysia
MES	Midsayap Experiment Station (PhilRice)
MORIF	Maros Research Institute for Food Crops, Sulawesi, Indonesia
NGOs	non-governmental organizations
NRCTP	National Rice Co-operative Testing Project, Philippines
ODA	Overseas Development Administration
ONs	observational nurseries
ORF	open reading frame
PAM-ELISA	parafilm mini-ELISA
PCR	polymerase chain reaction
PhilRice	Philippine Rice Research Institute
PSB	Philippine Seedboard
PYTs	preliminary yield trials
R&D	research and development
RIPA	rapid immuno filter paper assay
RT/PCR	reverse transcription/polymerase chain reaction
RTBV	rice tungro bacilliform virus
RTD	rice tungro disease
RTSV	rice tungro spherical virus
RTVD	rice tungro virus disease
RTVs	rice tungro viruses
TBS	tris-buffered saline
UPLB	University of the Philippines, Los Baños
WAI	weeks after inoculation
WAT	weeks after transplanting

Paper 1

Alternative diagnostic methods for detection of rice tungro viruses

P.Q. CABAUATAN and H. KOGANEZAWA*

International Rice Research Institute, PO Box 933, 1099 Manila, Philippines

INTRODUCTION

During the rice tungro virus disease (RTVD) workshop held in 1986 in South Sulawesi, Indonesia, five diagnostic techniques for rice tungro disease (RTD), commonly referred to as 'tungro', were presented, namely: symptomatology, transmission test, iodine/starch test, serological test and electron microscopy (Cabunagan *et al.*, 1986). Of these techniques, serological testing using enzyme-linked immunosorbent assay (ELISA) is the most commonly used for routine detection of the two rice tungro viruses (RTVs) in both epidemiological and resistance studies (Hibino *et al.*, 1987; Tiongco *et al.*, 1987). Conventional ELISA, however, is expensive, requires laboratory equipment that is not readily available in most laboratories in developing countries and cannot be done in the field. Also, ELISA cannot detect very low concentrations of virus as present in some rice varieties tolerant to rice tungro bacilliform virus (RTBV) (Takahashi *et al.*, 1993). The other techniques also have limitations, and symptomatology is not applicable to rice tungro spherical virus (RTSV), since it does not cause distinct symptoms in rice under tropical conditions. Thus, each diagnostic technique has limitations, and alternative techniques are required to overcome them. Since 1986, a number of diagnostic methods have been developed for plant virus detection. Some of these have been modified and applied to RTVs as alternative diagnostic methods. In this paper, four such methods are presented.

PARAFILM MINI-ELISA

Parafilm mini-ELISA (PAM-ELISA) is a modification of conventional ELISA. It involves the use of a parafilm membrane as the solid phase instead of microtitre plates (Tiongco *et al.*, 1992). A small piece of parafilm is wrapped around a glass microscope slide. Small imprints which serve as mini-wells are made by pressing one end of a glass rod (2 mm diameter) onto the membrane. Each well holds 10–15 µl of reagent. The procedure follows the principle of double-antibody sandwich-ELISA (DAS-ELISA) (Clark and Adams, 1977). After coating with specific antibody, the sample and conjugate (alkaline phosphatase-labelled specific antibody) are applied together. Nitroblue tetrazolium and 5-bromo-4-chloro-3-indolylphosphate are used as substrate. A blue colour reaction, indicating a positive result develops within 10 minutes. All procedures are done at room temperature and membranes are incubated for 3 h inside a covered plastic box lined with a moist paper towel to prevent drying. Washing of membranes after each step is done by gently flooding the membrane with wash buffer using either a pasteur pipette or a wash bottle. The technique can detect RTVs antigen in a 100-fold dilution of crude sap. Costs are considerably reduced and the assay time is shortened to 4 h by using this technique. PAM-ELISA does not require sophisticated equipment and can be readily adopted.

RAPID IMMUNOFILTER PAPER ASSAY

The rapid immunofilter paper assay (RIPA) was developed by Tsuda *et al.* (1992) and applied to detect plant viruses including cauliflower mosaic caulimovirus and tobacco mosaic tobamovirus. It involves the use of two kinds of latex beads (Japan Synthetic Rubber Co. Ltd, Tsukuba); white latex as the solid phase and pink latex as the tracer. The white and pink latexes are diluted separately with Tris-buffered saline (TBS = 0.02 M Tris, 0.15 M NaCl, pH 7.2) to 0.5 and 0.1% (v/v), respectively. The diluted latex beads are mixed with an equal volume of specific immunoglobulins (IgGs) at an antibody concentration of 100 µg/ml. The mixtures are incubated at room temperature for 3 h with continuous shaking and washed three times with TBS containing 0.1% bovine serum albumin (BSA) by centrifugation at 15 000 rpm for 10 min. The final pellets are resuspended in TBS-BSA at 0.5% and 0.1% (v/v) for the white and pink latex, respectively, and stored at 4°C.

The antibody-coated white latex is immobilized on Whatman glass filter paper strips (Whatman GF/A; 0.5 × 9 cm). A fine-pointed brush is used to apply a thin coat of white latex suspension on the filter paper strips *c.* 1.5 cm from the lower end. Latex-coated filter paper strips are air-dried and stored in a

*Present address: Shikoku National Agricultural Experiment Station, Ministry of Agriculture, Forestry and Fisheries, 1-3-1, Senyu, Zentsuji, Kagawa 765, Japan.

1

desiccator at room temperature until used. For virus assay, 0.1 g of infected rice leaf is homogenized in 900 l of extraction buffer (TBS with 0.01 Na_2SO_3) using a mortar and pestle. The extracts are clarified by centrifugation at 15 000 rpm for 10 min; 100 μl of clarified extract are placed in flat-bottomed Eppendorf tubes in which the lower end of the latex-coated filter paper strips is dipped until the extract is fully absorbed. Then, 100 μl of IgG-coated pink latex [diluted to 0.025% (v/v) with TBS] is added to the same tube. The pink latex suspension moves upward by capillary action and a pink band appears at the spot where the white latex was applied to indicate a positive reaction.

RIPA detects the virus antigen efficiently at different dilutions of the infected sap and there is no reaction with healthy sap. The sap dilution end-point depends on virus concentration in the plant and the titre of the antiserum. The sensitivity of RIPA is comparable to ELISA for both RTBV and RTSV. However, RIPA is simpler, less time-consuming and cheaper than ELISA. Moreover, RTBV and RTSV can be detected simultaneously in a single filter paper strip sensitized with antibody to both viruses. RIPA is as simple as using pH test paper strips. Latex-coated filter paper strips can be prepared in advance and stored in a desiccator for about six months. For detection of RTBV and RTSV, current RIPA requires the centrifugation of plant extracts, but it can be modified for field application; for example by using organic solvents to clarify the extracts.

POLYMERASE CHAIN REACTION

Polymerase chain reaction (PCR) involves the *in vitro* amplification of nucleic acid sequences (Saiki *et al.*, 1988). It has been applied to detect small amounts of RTBV in the leafhopper vector (*Nephotettix virescens*) and in rice cultivars tolerant to RTBV (Takahashi *et al.*, 1993). For PCR, DNA is dissolved in water after extraction from the infected tissue with phenol. The PCR products are then subjected to electrophoresis in 1% agarose gel. The gels are stained with ethidium bromide (0.5 μg/ml buffer). PCR is $10^3 - 10^4$ more sensitive than ELISA. Many plants of rice cultivars Utri Merah, Utri Rajapan and Balimau Putih infected with RTBV reacted negatively in ELISA but were positive for RTBV in PCR (Takahashi *et al.*, 1993). PCR can be used as a supplement in evaluating RTBV-tolerant cultivars which support very low virus concentrations (Koganezawa and Cabunagan, page 54).

INDICATOR PLANTS FOR RTSV DETECTION

Epidemiological studies of RTD may involve the monitoring of viruliferous leafhoppers caught in the field by infectivity assays using susceptible cultivars. Because two viruses are involved in tungro disease and since RTSV does not cause any distinct symptom on infected plants, it is necessary to index the inoculated plants serologically. A suitable indicator plant is an economical and convenient alternative.

During a routine screening for tungro resistance at the International Rice Research Institute (IRRI), an accession of *Oryza glaberrima* (IRGC 100139) showed unusually severe necrosis when infected with RTVs. Few plants survived more than three weeks after inoculation. IRGC 100139 showed distinct symptoms when infected with RTSV alone. Infected plants were markedly stunted with pale green, narrow and erect leaves about three weeks after inoculation (Cabauatan *et al.*, 1993). This is the first accession tested at IRRI that shows distinct symptoms of RTSV infection. Visual evaluation of RTSV-infected IRGC 100139 plants corresponded well with serological assay (Table 1). Symptom severity and survival of IRGC 100139 plants correlated well with infection with either RTBV alone or both RTBV and RTSV. Generally, infected plants that survived more than three weeks after inoculation were infected with RTBV alone, while those that did not were infected with both RTBV and RTSV. Susceptibility of IRGC 100139 to virus and vector and its suitability as a virus source were comparable to the susceptible standard Taichung Native 1 (TN1) (Tables 2 and 3).

Table 1 Percentage of RTSV infection on inoculations to *Oryza glaberrima* (IRGC 100139) and rice cv. TN1

Variety	No. plants inoculated	% plants infected	
		Visual	ELISA
O. glaberrima	220	90.0	90.5
TN1	200	0.0	86.5

Note: Inoculated seven days after sowing and assessed visually one month later and also by ELISA.

Table 2   Transmission rate of RTSV to and from *O. glaberrima* (IRGC 100139)and rice cv. TN1

Virus source	TN1	*O. glaberrima*
O. glaberrima	85.7 (66/77)	90.0 (72/80)
TN1	88.7 (71/80)	90.0 (72/80)

Notes: Inoculated seven days after sowing and tested one month later.
Figures in parentheses represent the number of infectd plants over the number inoculated.

Table 3   Antibiosis tests with *O. glaberrima*, TN1 and ARC 11554 to green leafhopper

Variety (IRGC Acc. No.)	Nymph survival rate
O. glaberrima (IRGC 100139)	82.3
"	64.0
"	56.0
"	76.7
TN1	95.0
ARC 11554	32.0

Note: The weighted mean of nymph survival rate was calculated from the following formula:

$$\text{Weighted mean} = \frac{A_1 + A_2 \times 2 + A_3 \times 3}{1 + 2 + 3}$$

where A_1, A_2 and A_3 represent the average survival rate (%) 1, 2 and 3 days after caging, respectively on 10-day-old seedlings.

IRGC 100139 has the attributes of an ideal indicator host for RTVs as it is susceptible to the vector and becomes a good virus source when infected. The use of IRGC 100139 as an indicator host would be a cheap and reliable alternative to ELISA in epidemiological studies of RTD without the need for expensive serological assays.

SUMMARY

The alternative diagnostic techniques for detecting RTVs can be used to supplement existing methods. The choice of alternative technique depends on the objective and/or nature of the research being done and on the availability of materials and equipment. For example, in screening hundreds of rice accessions for resistance to RTVs, the conventional DAS-ELISA is preferable. For rapid testing a few samples either in the laboratory or in the field, PAM-ELISA or RIPA can be used. By contrast, indicator plants can be used to monitor RTSV-carrying leafhoppers without the need for serological assay. Lastly, PCR can be applied to detect low concentrations of RTBV in varieties with tolerance to the virus and as a supplement in evaluating varietal resistance to RTBV.

REFERENCES

CABAUATAN, P.Q., KOBAYASHI, N., IKEDA, R. and KOGANEZAWA, H. (1993) *Oryza glaberrima:* an indicator plant for rice tungro spherical virus. *International Journal of Pest Management,* **39**: 273–276.

CABUNAGAN, R.C., FLORES, Z.M., TIONGCO, E.R. and HIBINO, H. (1986) Diagnostic techniques for rice tungro disease. pp. 1317. In: *Proceedings of the Workshop on Rice Tungro Virus. 24–27 September 1986, Maros, South Sulawesi, Indonesia.* Indonesia: Ministry of Agriculture.

CLARK, M.F. and ADAMS, A.N. (1977) Characteristics of the microplate method of enzyme-linked immunosorbent assay for the detection of plant viruses. *Journal of General Virology,* **34**: 475–483.

HIBINO, H., TIONGCO, E.R., CABUNAGAN, R.C. and FLORES, Z.M. (1987) Resistance to rice tungro-associated viruses in rice under experimental and natural conditions. *Phytopathology,* **77**: 871–875.

SAIKI, R.K., GELFAND, D.H., STOFFEL, S., SCHARF, S.J., HIGUCHI, R.G., HORN, G.T., MULLIS, K.G. and ERLICH, H.A. (1988) Primer-directed enzymatic amplification of DNA with a thermostable DNA polymerase. *Science,* **239**: 487–491.

TAKAHASHI, Y., TIONGCO, E.R., CABAUATAN, P.Q., KOGANEZAWA, H., HIBINO, H. and OMURA, T. (1993) Detection of rice tungro bacilliform virus by polymerase chain reaction for assessing mild infection of plants and viruliferous vector leafhoppers. *Phytopathology,* **83**: 655–659.

TIONGCO, E.R., CABUNAGAN, R.C., FLORES, Z.M. and HIBINO, H. (1987) Critical time for the occurrence and development of tungro infection in the field. *Transactions of the National Academy of Science and Technology (Philippines),* **9**: 433–441.

TIONGCO, E.R., FLORES, Z.M., KOGANEZAWA, H. and TENG, P.S. (1992) Parafilm membrane for sero-assay of rice viruses. Paper presented at the *Twenty-ninth Annual Meeting of the Philippine Phytopathological Society, 27–30 April 1992, Tagaytay City, Philippines.*

TSUDA, S., KAMEYA-IWAKI, M., HANADA, K., KOUDA, Y., HIKATA, M. and TOMARU, K. (1992) A novel detection and identification technique for plant viruses: Rapid immunofilter paper assay (RIPA). *Plant Disease,* **76**: 466–469.

Paper 2

Molecular biology of variation in rice tungro viruses

R. HULL, J.M. HAY, A. DRUKA, Z. FAN, V. THOLE and S. ZHANG

John Innes Centre, Norwich Research Park, Colney, Norwich NR4 7UH, UK

INTRODUCTION

Tungro-like diseases have been recognized in rice for over a century, being first described in Indonesia in the 1850s. The disease is endemic in many rice-growing countries of South and South-East Asia (Figure 1) and its prevalence over the last two decades can be attributed to the increase of well-irrigated areas and the wide cultivation of rice varieties with similar genetic background. One of the characteristics of tungro epidemics is that they come and go suddenly (Hibino, 1987). However, tungro is considered by rice growers to be of major significance, as is shown by the large number of local names used for the disease.

Rice tungro virus disease, commonly referred to as tungro or rice tungro disease, is caused by a combination of two viruses: rice tungro spherical virus (RTSV) and rice tungro bacilliform badnavirus (RTBV) (Saito *et al.*, 1976; Ou, 1985; Hibino, 1987). RTSV causes few symptoms and is transmitted in the semi-persistent manner by several leafhopper species, primarily by the green leafhopper (GLH) *Nephotettix virescens*. RTBV alone causes severe symptoms, but is not transmitted in the absence of RTSV. In joint infections RTBV is transmitted by the GLH and it is considered that RTSV provides some factor which assists the transmission (Cabauatan and Hibino, 1985; Hibino, 1987). Thus, in the combined disease RTSV provides the transmission and RTBV most of the symptoms.

There has been much work recently on the molecular biology of the two tungro viruses. RTSV contains a single-stranded polyadenylated RNA of about 12.2 kb as its genome (Shen *et al.*, 1993; Thole *et al.*, unpublished observation). The sequence of this RNA shows that it has a large open reading frame (ORF) capable of encoding a protein of 390 kDa and two short ORFs at the 3' end (see Figure 2) (Shen *et al.*, 1993). The assignation of NTP, Pro and Pol domains is based on amino acid homologies with gene products from other viruses. It is thought that the large protein is a polyprotein which is cleaved to its functional units by one domain which has homology to a cysteine protease. Among the products from this polyprotein are the three-coat protein species which make up the virus capsid; the N-terminus of each of the coat proteins has been determined (Zhang *et al.*, 1993; Shen *et al.*, 1993) and thus they can be mapped onto the viral genome.

RTBV is a pararetrovirus with a circular double-stranded DNA genome of 8.0 kbp (Jones *et al.*, 1991; Hay *et al.*, 1991; Qu *et al.*, 1991; Hibino *et al.*, 1991). The nucleotide sequences of three isolates of RTBV from the Philippines (Hay *et al.*, 1991; Qu *et al.*, 1991; Kano *et al.*, 1992) have been reported and show *c.* 98% similarity with each other with only scattered nucleotide substitutions. The sequence suggests that RTBV contains four ORFs encoding proteins of 24 kDa, 12 kDa, 194 kDa and 46 kDa (see Figure 3). Functions are not yet known for the three smaller products though they have been identified in infected plants (Hay *et al.*, 1994; Lee *et al.*, personal communication). It appears that the product of the largest ORF is a polyprotein which is processed by an aspartate protease domain into several products including the viral coat protein and the polymerase (reverse transcriptase and ribonuclease H).

There have been many attempts to breed resistance to the tungro viruses, but little success in finding durable natural resistance genes. The developments over the last decade, leading to the concept of non-conventional protection (for recent review see Hull, 1994) by viral sequences introduced into the plant genome, have provided the stimulus for testing this approach for control of tungro. As well as the detailed molecular information required in the design of the transgenes, it is necessary to have an understanding of the variation of the viruses over the whole region in which the disease is endemic. This can then be used in identifying conserved regions of the genomes for targeting which should ensure more durable protection.

This paper reviews the information and understanding of the variation of tungro viruses up to 1994.

BIOLOGICAL INFORMATION ON VARIATION

Several strains have been reported for tungro based on different symptoms and reactions in certain rice cultivars. Mild and severe strains were described from the Philippines (Rivera and Ou, 1967) and five strains were designated in India (Shastry *et al.*, 1972; Anjaneyulu and John, 1972; Mishra *et al.*, 1976;

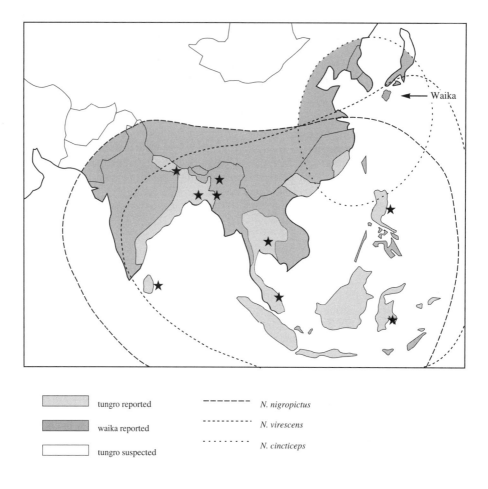

	tungro reported	– – – – – – –	*N. nigropictus*
	waika reported	– · – · – · –	*N. virescens*
	tungro suspected	· · · · · · ·	*N. cincticeps*

Figure 1   Geographic distribution of rice tungro and waika diseases and various green leafhopper (*Nephotettix*) species; the stars indicate sites from which isolates of tungro were obtained.

Basu *et al.*, 1976). However, these were described before it was recognized that tungro was caused by a combination of two viruses and it was unknown as to which of the viruses was varying. Recently, symptom variants were separated from a field isolate of RTBV from the Philippines (Cabauatan and Koganezawa, 1994) and also a strain of RTSV has been reported to overcome the resistance of variety TKM6 (Cabauatan *et al.*, 1994). Apart from a restriction fragment length analysis of the DNA of the symptom variants of RTBV (Dolores-Talens *et al.*, 1994), there is no molecular characterization of these strains. Also, because of quarantine regulations it was not possible to compare strains from different countries.

MOLECULAR INFORMATION ON VARIATION

As part of a project to develop non-conventional protection against the tungro viruses, isolates of the disease were collected from Bangladesh, India (West Bengal, Orissa and Assam), Indonesia (Bali and South Sulawesi), Malaysia, Nepal, Philippines, Sri Lanka and Thailand (Figure 1) and studied at the John Innes Centre. Details of these studies will be reported elsewhere — a summary of the findings is presented here.

Rice tungro spherical virus

The study of the variation of this large RNA genome has focused on the region encoding the three coat protein species (Figure 2). Using primers designed to cover the 2.1 kb coding region of all three coat proteins, reverse transcription/polymerase chain reaction (RT/PCR) products were obtained from RNA extracted from isolates from Bangladesh, India, Malaysia, Philippines and Thailand. Aliquots of these PCR products were electrophoresed in gels, blotted onto membranes and probed with the PCR products which had been labelled with [32]P. Autoradiography showed that the Philippine and Malaysian probes hybridized with the PCR product from the Philippine, Malaysian and Thai isolates; the Indian

RTSV   12433 nucleotides

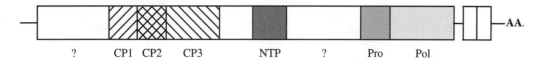

CP   = coat protein
NTP = nucleotide triphosphate binding domain
Pro = protease domain
Pol = RNA-dependent RNA polymerase domain
?   = domains with no ascribable function
☐   = non-coding 5´ and 3´sequences
AAA = poly-adenylate

Figure 2   Genome organization of RTSV (after Shen *et al.*, 1993). The boxes indicate the open reading frames (ORFs) with the functional domain. The assignation of NTP, Pro and Pol domains is based on amino acid homologies with gene products from other viruses.

probe hybridized with the Indian and Thai isolates and the Bangladesh probe hybridized with the Bangladesh and Thai isolates. Thus, as far as the coat protein coding region is concerned, there appear to be three variants, one from the Philippines and Malaysia, another from India and the third from Bangladesh; the Thai isolate appears to be a mixture of all three. This experiment shows the power of examining the populations revealed by using all the PCR products as these differences may not have been revealed if the products had been cloned. Even with this approach any very different variants which were not compatible with the PCR primers would not have been recognized. There still remains the need to determine the true extent of the variation in this coat protein region and also to assess the variation in other regions of the RTSV genome.

Antisera have been raised against each of the three RTSV coat protein species of the Philippine isolate expressed as fusion proteins in *Escherichia coli*. These react specifically with the target coat protein species. Western blots of the coat proteins 1 and 2 of both the Philippine and Indian isolates either from purified virus or crude sap give bands of the expected sizes. The coat protein 3 in purified preparations of the Philippine isolate is mainly of the expected size of 33 kDa with a small amount of larger material (45–49 kDa); in crude sap the 33 kDa band is not found and all the reacting material is in the larger bands. Only the larger bands are found in both purified virus and crude sap of the Indian isolate indicating that there might be differences in sequence at the amino acid level as well as at the nucleic acid level (A. Druka, unpublished observation).

Rice tungro bacilliform virus

The DNA genomes of isolates of RTBV from Bangladesh, India (West Bengal and Orissa), Indonesia (South Sulawesi), Malaysia and Thailand as well as the type Philippine isolate have been cloned in bacterial plasmids, and cross-hybridization experiments were carried out on the RTBV moiety. These isolates fell into two groups: one based on the Indian subcontinent (Indian and Bangladesh isolates) and the other on the Association of South East Asian Nations (ASEAN) trading states (Indonesia, Malaysia, Philippines and Thailand). Within each group the DNA cross-hybridized strongly, whereas between the groups there was only weak hybridization. DNA preparations from infected leaves showed the isolates from Assam, Nepal and Sri Lanka fell into the Indian group whereas isolates from Sabah (Malaysia) and Bali (Indonesia) belonged to the ASEAN group. Furthermore, the DNAs of the ASEAN isolates had single sites for *Sal*I and *Bam*HI restriction enzymes whereas those from the Indian isolates lacked the *Sal*I site and had the *Bam*HI site in a different place.

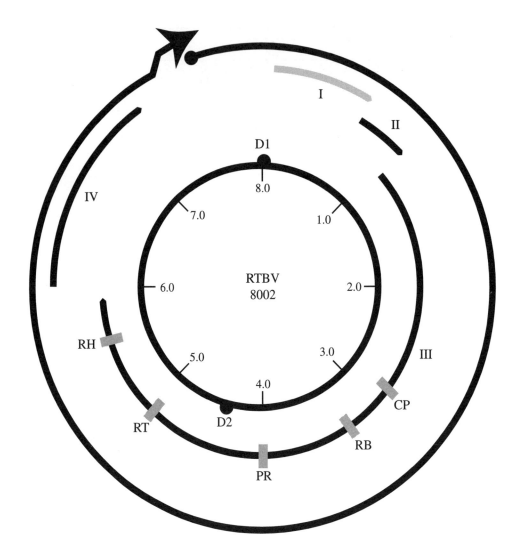

CP = suggested N-terminus of 33KDa coat protein species
RB = RNA binding consensus sequence
PR = aspartate protease domain
RT = reverse transcriptase domain
RH = ribonuclease H domain
D1 = discontinuity
D2 = discontinuity

Figure 3 Genome organization of RTBV (after Hay *et al.*, 1991; Qu *et al.*, 1991). The inner circle represents the double-stranded DNA genome with the positions of the discontinuities being marked D1 and D2 and the nucleotides being marked in kbp; the arcs represent the four open reading frames (I–IV) with the putative functions based on amino acid sequence homology indicated. The outer circle represents the more than genome length RNA transcript.

The part of the Indian (West Bengal) isolate from the C-terminus of ORF III to the N-terminus of ORF III via the discontinuity 1 (Figure 3) has been sequenced. Comparison of this sequence with that of the equivalent region of the Philippine isolate showed *c.* 60–70% homology. Sequencing also showed that there was a deletion of *c.* 60 nucleotides in the Indian isolate when compared with the

Philippine isolate at *c*. nucleotide 7950 (Fan, unpublished observation). PCR primers were made to conserved sequences on either side of this deletion which would give a product of 290 nucleotides for the Philippine isolate and of 230 nucleotides for the Indian isolate. PCR amplifications showed that this deletion was not the effect of cloning or of glasshouse culture of the Indian isolate and that it was found in all isolates tested from the Indian subcontinent; there was no evidence of the deletion in any of the isolates of RTBV tested from the ASEAN states.

CONCLUSIONS

This molecular analysis has revealed variation in both the tungro viruses; that of RTBV appears to be marked and geographically delimited. On the present evidence there appear to be two strains of RTBV, one in the Indian subcontinent and the other in the ASEAN states. The geographic distribution of the variants of RTSV based on the coat protein region (less than 20% of the genome) is not so clearly defined. These differences raise several interesting points. Firstly, it appears that the epidemiology of the two viruses is not closely interlinked. Secondly, the two strains of RTBV are separated by a climatic divider along the Bay of Bengal and through Myanmar which would suggest that they have been separated for some considerable time.

It is not possible to assess the significance of this variation on the protection offered by non-conventional transgenes until transgenic plants become available. However, an understanding of the variation is important in the design of tests which would indicate the durability of such forms of resistance.

ACKNOWLEDGEMENTS

This work was supported by grants from the Rockefeller Foundation, the Natural Resources Institute and the European Community Science and Technology for Development Programme. The import of tungro viruses and their maintenance and genetic manipulation were covered by various licences from the Ministry of Agriculture, Fisheries and Food.

REFERENCES

ANJANEYULU, A. and JOHN, V.T. (1972) Strains of rice tungro virus. *Phytopathology,* **62**: 1116–1119.

BASU, A.N., MISHRA, M.D., NIAZI, F.R. and GHOSH, A. (1976) A proposed key to the strains of rice tungro virus. *International Rice Research Newsletter,* **1**: 6–7.

CABAUATAN, P.Q. and HIBINO, H. (1985) Transmission of rice tungro bacilliform and spherical viruses by *Nephotettix virescens* Distant. *Philippines Phytopathology,* **21**: 103–109.

CABAUATAN, P.Q. and KOGANEZAWA, H. (1994) Symptomatic strains of rice tungro bacilliform virus. *International Rice Research Notes,* **19**: 11–12.

CABAUATAN, P.Q., CABUNAGAN, R.C. and KOGANEZAWA, H. (1994) Comparative transmission of two strains of rice tungro spherical virus in the Philippines. *International Rice Research Notes,* **19**(2): 10–11.

DOLORES-TALENS, A.C., ESCARA-WILKIE, J.R., CABAUATAN, P.Q., NELSON, R.J. and KOGANEZAWA, H. (1994) Strain differentiation of rice tungro bacilliform virus by restriction fragment length analysis of polymerase chain reaction-amplified products. *International Rice Research Notes,* **19**(1): 10–11.

HAY, J.M., JONES, M.C., BLAKEBROUGH, M.L., DASGUPTA, I., DAVIES, J.W. and HULL, R. (1991) An analysis of the sequence of an infectious clone of rice tungro bacilliform virus, a plant pararetrovirus. *Nucleic Acids Research,* **19**: 2615–2621.

HAY, J., GRIECO, F., DRUKA, A., PINNER, M., LEE, S.-C. and HULL, R. (1994) Detection of rice tungro bacilliform virus gene products *in vivo. Virology,* **205**: 430–437.

HIBINO, H. (1987) Rice tungro virus disease: current research and prospects. In: *Proceedings of the Workshop on Rice Tungro Virus.* Indonesia: Ministry of Agriculture.

HIBINO, H., ISHIKAWA, K., OMURA, T., CABAUATAN, P.Q. and KOGANEZAWA, H. (1991) Characterization of rice tungro bacilliform and rice tungro spherical virus. *Phytopathology,* **81**: 1130–1132.

HULL, R. (1994) Resistance to plant viruses: obtaining genes by non-conventional approaches. *Euphytica,* **75**: 195–205.

JONES, M.C., GROUGH, K., DASGUPTA, I., SUBBA RAO, B.L., CLIFFE, J., QU, R., SHEN, P., KANIEWSKA, M., BLAKEBROUGH, M., DAVIES, J.W., BEACHY, R.N. and HULL, R. (1991) Rice tungro disease is caused by an RNA and a DNA virus. *Journal of General Virology,* **72**: 757–761.

KANO, H., KOIZUMI, M., NODA, H., HIBINO, H., ISHIKAWA, K., OMURA, T., CABAUATAN, P.Q. and KOGANEZAWA, H. (1992) Nucleotide sequence of capsid protein gene of rice tungro bacilliform virus. *Archives of Virology,* **124**: 157–163.

MISHRA, M.D., NIAZI, F.R., BASU, A.N., GHOSH, A. and RAYCHAUDHURI, S.P. (1976) Detection and characterization of a strain of rice tungro virus in India. *Plant Disease Reporter,* **60**: 23–25.

OU, S.H. (1985) *Rice Diseases.* 2nd edn. Kew, UK: Commonwealth Mycological Institute.

QU, R., BHATTACHARYYA, M., LACO, G.S., DE KOCHKO, A., SUBBA RAO, B.L., KANIEWSKA, M. B., ELMER, J.S., ROCHESTER, D.E., SMITH, C.E. and BEACHY, R.N. (1991) Characterization of the genome of rice tungro bacilliform virus: comparison with Commelina yellow mottle virus and caulimoviruses. *Virology,* **185**: 354–364.

RIVERA, C.T. and OU, S.H. (1967) Transmission studies on two strains of rice tungro virus. *Plant Disease Reporter,* **51**: 877–881.

SAITO, Y., IWAKI, M. and USINGI, T. (1976) Association of two types of particle with tungro-group disease of rice. *Annals of the Phytopathological Society of Japan,* **43**: 375.

SHASTRY, S.V.S., JOHN, V.T. and SESHU, D.V. (1972) Breeding for resistance to rice tungro virus in India. pp. 239–252. In: *Plant Breeding.* Los Baños, Philippines: International Rice Research Institute.

SHEN, P., KANIEWSKA, M., SMITH, C. and BEACHY, R.N. (1993) Nucleotide sequence and genomic organisation of rice tungro spherical virus. *Virology,* **193**: 621–630.

ZHANG, S., JONES, M.C., BARKER, P., DAVIES, J.W. and HULL, R. (1993) Molecular cloning and sequencing of coat protein-encoding cDNA of rice tungro spherical virus — a plant picornavirus. *Virus Genes,* **7**: 121–132.

Paper 3

Effect of introduced sources of inoculum on tungro disease spread in different rice varieties

M.K. SATAPATHY[1]*, T.C.B. CHANCELLOR[2], P.S. TENG[1], E.R. TIONGCO[1]† and J.M. THRESH[2]

[1]*International Rice Research Institute, PO Box 933, 1099 Manila, Philippines*
[2]*Natural Resources Institute, Chatham Maritime, Chatham, Kent ME4 4TB, UK*

INTRODUCTION

Much information is available on the aetiology of rice tungro virus disease, also referred to as tungro, and on its transmission characteristics (Hibino and Cabunagan, 1986) but considerably less is known about its epidemiology. The limited understanding of tungro dynamics has hindered the development of coherent strategies to manage the disease (Thresh, 1989), although recent studies conducted in Indonesia have provided significant advances in improving tungro forecasting in asynchronously cropped rice areas (Suzuki *et al.*, 1992; Suzuki *et al.*, page 30). An analysis of historical survey data in the Philippines has thrown new light on factors influencing tungro incidence in endemic and non-endemic areas, particularly with regard to the relative importance of leafhopper abundance and the proportion of vectors which are infective (Savary *et al.*, 1993). Nevertheless, many gaps remain in our knowledge of tungro epidemiology.

One important area where information is lacking is on the pattern, sequence and distance of tungro spread into and within rice plantings (Thresh, 1989). There is some evidence for the occurrence of tungro disease gradients (Kondaiah *et al.*, 1976) but the role and relative importance of tungro spread into and within plantings is not known. In order to obtain more information, a series of field experiments was designed to examine the role of inoculum sources in tungro disease spread. This paper summarizes the main findings from these experiments, conducted from 1991 to 1993 in the Philippines, and discusses the methodologies developed and their application for epidemiological studies. The first trial was designed to assess the effect of introduced sources on tungro disease spread and to quantify disease progress in varieties with different degrees of resistance to the main vector, the green leafhopper *Nephotettix virescens*, and to tungro viruses. The objectives of the second and third trials were to assess the effects of introduced inoculum sources of different strengths and spatial distribution, respectively, on tungro disease spread.

EXPERIMENTAL DETAILS

The experiments were conducted on the experimental farm of the International Rice Research Institute (IRRI) in Los Baños, Laguna Province. The approach adopted was to record tungro incidence in replicated field plots and to measure disease spread along transects from introduced sources of infection. The optimum plot size, shape and orientation and the required separation distance between plots were not known before the start of the experiments. In addition, it was difficult to estimate the size of the introduced sources required to ensure that adequate spread occurred. It was recognized that the potential for inter-plot interference needed to be considered in the experimental design, as well as the possible effects of exogenous inoculum. Inoculum sources are present on the IRRI farm throughout most of the year and include inoculum deliberately introduced into breeders' plots for routine screening of germplasm for tungro resistance. Further inoculum sources result from the many plantings of susceptible varieties and are maintained by the continuous cultivation of rice in certain areas of the farm throughout the year. Care was taken to ensure that the experiments were located away from any obvious sources of infection. In one experiment where tungro was observed to be present in an area located *c.* 50 m from the trial site, the blocks were oriented in such a way that the potential exposure to exogenous inoculum was similar for each treatment.

Tungro spread in different varieties with and without introduced inoculum sources

In the 1991 wet season (WS91) experiment five varieties were selected for use, based on their differing degrees of resistance to *N. virescens* and to tungro disease at the time they were released:

---

*Present address: Regional College of Education, Bhubaneswar 751007, Orissa, India.
†Present address: Philippine Rice Research Institute, Maligaya, Nueva Ecija, Philippines.

11

- IR22 was classified as susceptible to *N. virescens* and to tungro disease
- IR26, when released, was regarded as being resistant to *N. virescens* and field-resistant to tungro disease; the field-resistance was subsequently shown to be due to its resistance to rice tungro spherical virus (RTSV) but not to rice tungro bacilliform virus (RTBV) (Hibino *et al.*, 1988)
- IR36 was classified as moderately resistant to *N. virescens* and field-resistant to tungro disease
- IR54 was classified as resistant to *N. virescens* and field-resistant to tungro disease
- IR72 was classified as resistant to *N. virescens* and field-resistant to tungro disease.

The five varieties were assigned to the main plots of a randomized block design with three replicates. Each replicate was located in a different part of the farm and was surrounded by other rice fields and experimental plots. Each main plot was divided into sub-plots, one with and one without a central introduced source of tungro-infected plants. Each sub-plot was 19 × 19 m and there were uncultivated strips 3 m wide between sub-plots and 0.5 m wide between main plots. Seedlings were raised in an insect-proof screenhouse and transplanted manually at 20 days after transplanting (DAT) on 12 July at a 20 × 20 cm spacing, using three or four seedlings per hill. The source comprised a 10 × 10 array of 100 infected hills in the centre of one sub-plot of each main plot, at the standard spacing (Figure 1). The plants in the source areas were the same age as the surrounding ones. The equivalent area of each sub-plot without an inoculum source was planted with uninfected seedlings of the appropriate variety.

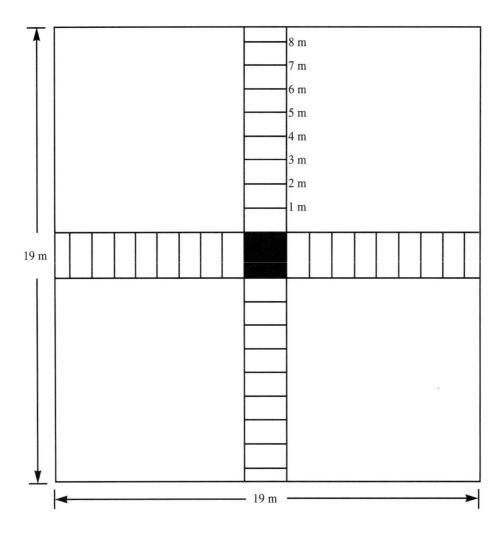

Figure 1   Schematic diagram of a single sub-plot in the 1991 wet season trial showing in solid black the centrally located 2 × 2 m area containing 100 rice hills (inoculated with tungro viruses or uninoculated, according to the 'source' treatment; see text for details) and the four arms of the cruciform sampling area.

Disease incidence was assessed only on the 1600 hills within a cruciform shaped sample area 2 m wide running through the centre of each sub-plot (Figure 1). Hills with tungro symptoms were counted weekly in each 2 × 1 m section of the sample area starting 13 DAT. Gradients of disease were based on the overall means for the incidence of infection in the twelve 2 × 1 m areas recorded for each variety at each of eight distances from the central source. This approach was adopted as a means of smoothing out irregularities in the data for individual arms of the cruciform areas recorded and because preliminary analyses indicated that spread did not differ greatly in each direction from the central sources.

Populations of the major leafhopper vectors of tungro disease, *Nephotettix virescens* (Distant), *N. nigropictus* (Stål) and *Recilia dorsalis* (Motschulsky), were assessed weekly using an insect net of 30 cm diameter. In order to reduce disturbance, the areas sampled excluded those where weekly recording of tungro symptoms was conducted. Each of the four sections outside the cruciform sampling area was sampled in sequence in a 4-week cycle starting 10 DAT (Figure 1). Ten continuous sweeps were made in each sub-plot and the leafhoppers caught were killed with ethyl acetate and taken to the laboratory for identification and counting. Standard agronomic practices were followed, including hand weeding; no pesticides were applied at any stage of the experiment. Full details of cultural practices, raising of seedlings, inoculation of source plants and serological testing of leaf samples for confirmation of the presence of tungro are given by Satapathy *et al.* (1996).

In the 1992 dry season experiment (DS92), which was planted on 29 January, variety IR54 was omitted because it gave results similar to those of IR22 in the WS91 experiment. As results for IR54 in WS91 were similar to those for IR22 in terms of disease incidence and leafhopper abundance, they are not presented here. There were four replications of each treatment and all four blocks were located together at one site. The plot size was increased to 22 × 22 m, as disease had spread to the plot edges in the susceptible varieties by the end of the WS91 trial.

Effect of source size on tungro disease spread

Field experiments were conducted in the 1992 wet season (WS92) and the 1993 dry season (DS93) in which five source treatments of 0, 1, 5, 25 and 125 tungro-infected hills, respectively, were planted in the centre of each plot. A randomized complete block design was used with four replications. Plots were 22 × 22 m in size, with an uncultivated area *c.* 2 m wide between plots within a block and between blocks. In WS92, a susceptible variety, IR22, and a moderately resistant variety, IR36, were used. In DS93, only IR22 was used as results from the first set of trials suggested that disease incidence was much lower in the dry season. Planting was on 7 August and 3 February in the WS92 and DS93 experiments, respectively. Cultural practices, methods for raising infected seedlings and recording procedures for tungro disease incidence and leafhopper number were similar to those used in the WS91 and DS92 experiments.

Effect of the disposition of sources on tungro disease incidence

Field trials were carried out in the 1993 dry season (DS93) and wet season (WS93), in which the spatial distribution of 100 tungro-infected hills was varied in four treatments (Figure 2):

- a single block of 10 × 10 infected hills in the centre of the plot
- 25 blocks of 2 × 2 infected hills distributed uniformly throughout the plot
- 100 single infected hill sources distributed uniformly throughout the plot
- no infected hills.

A randomized complete block design was used with one variety, IR22, and four replicates. The plot size was 20 × 20 m with uncultivated areas of 5 m wide between plots within a block and between blocks. Planting was on 3 February and 14 July in DS93 and WS93, respectively. The total number of infected hills was recorded weekly in one quarter of each plot and the disease status of each hill recorded in one replicate of each treatment. Leafhopper population estimates were obtained by sweep-net counts from one of each of the remaining three quarters of the plot in weekly rotation. Cultural practices and methods of raising infected seedlings were similar to those used in the earlier experiments.

13

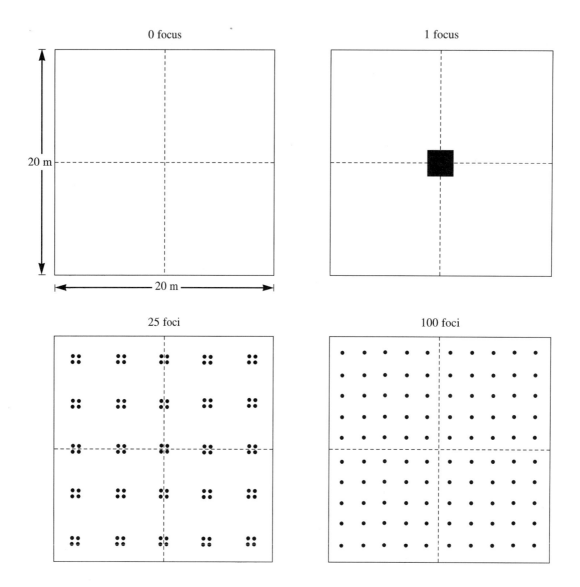

Figure 2   Distribution of foci of infection for the four source treatments (0, 1, 25 and 100 foci, respectively) in the 1993 wet and dry season source disposition experiments.

Statistical analysis

In the WS92 and DS93 source size experiments, and in the DS93 and WS93 source disposition experiments, analysis of variance was carried out on final tungro disease incidence. Arc sine-transformed values were used in the analyses to stabilize the variances. Comparison of mean values among varieties was made by Duncan's multiple range test.

RESULTS

Effects of introduced inoculum sources on tungro incidence

In all the experiments, disease increase began earlier and final tungro incidence was greater in sub-plots or plots with introduced sources of inoculum than in those without. For example, in the WS91 experiment mean tungro disease incidence at 34 DAT for all varieties was 12% and 1% for sub-plots with and without an introduced source, respectively (Figure 3). By 69 DAT, when the last disease recordings were taken, tungro incidence in the source and no-source treatments had increased to 43% and 12%, respectively. The same pattern was observed in the DS92 experiment when tungro incidence was much lower (Figure 3).

There were large differences between varieties in the final incidence of disease in these two experiments, but the effects of the introduced sources were apparent for all varieties. In the WS91 experiment, tungro incidence was greatest in IR22, reaching 83% at 69 DAT in sub-plots with introduced sources, compared with 34% in sub-plots without sources (Figure 4). Tungro incidence was lowest in IR72, reaching 2% in sub-plots with introduced sources and 0.1% in

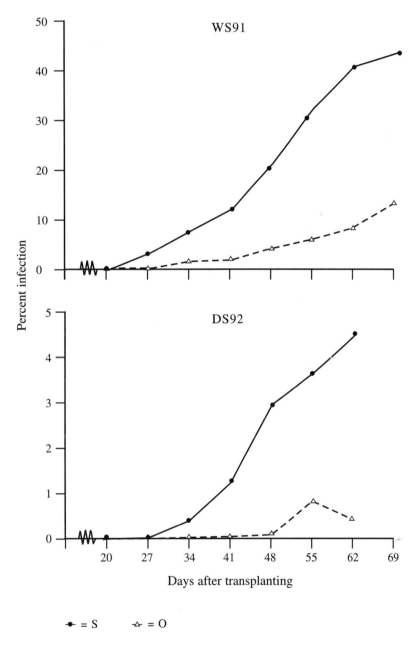

Figure 3   Tungro disease progress curves, based on mean values for four varieties, with (S) and without (O) introduced sources of inoculum in the 1991 wet season (WS91) and the 1992 dry season (DS92) experiments. Note the difference in the vertical scale.

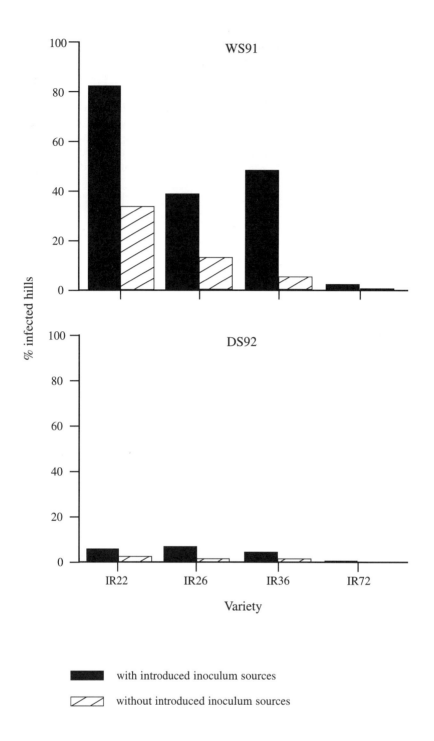

Figure 4 Final tungro disease incidence 69 DAT in four rice varieties with and without introduced sources of inoculum in the 1991 wet season (WS91) and the 1992 dry season (DS92) experiments.

sub-plots without sources. Intermediate levels of disease occurred in IR26 and IR36 in the WS91 experiment, but incidence was similar to that in IR22 in DS92 when disease levels were much lower (Figure 4).

The **size** of the introduced inoculum sources affected tungro disease incidence in both the WS92 and DS93 source size experiments, as illustrated by the results for IR22. However in WS92 final incidence in plots with 1- and 5-hill sources was similar to that in the control plots with no sources (Table 1). In DS93, when disease levels were lower, the effect of the single-hill source was apparent with final incidence reaching 3%, significantly higher ($p<0.05$) than the value of <1% for plots without sources.

Table 1 Final tungro disease incidence (%) in relation to source size in rice variety IR22 in the 1992 wet season (WS92) and 1993 dry season (DS93) experiments

Source size (no. of hills)	WS92	DS93
0	35.6a	0.6a
1	36.6a	3.3b
5	40.7a	3.4b
25	51.4b	11.0c
125	76.2c	17.5d

Note: In a column, means (average of four replications) followed by a common letter are not significantly different at the 5% level by Duncan's multiple range test. Values were transformed using the arc sine transformation for the analysis.

In both experiments, tungro incidence in plots with 25-hill sources was significantly greater ($p<0.05$) than in plots with 5-hill sources. Similarly, incidence in plots with 125-hill sources was significantly greater ($p<0.05$) than in plots with 25-hill sources.

Tungro incidence was influenced by the **disposition** of the introduced inoculum sources. Unlike in the two previous sets of experiments, disease levels were similar in the wet and dry seasons. In the DS93 source disposition experiment, final tungro disease incidence was significantly greater ($p<0.05$) in both the 25- and 100-foci treatments than in the single-focus treatment (Table 2). In the WS93 source disposition experiment, final incidence in the plots with 100 introduced disease foci was significantly greater ($p<0.05$) than that in the single-foci plots, but not the 25-foci plots.

Disease gradients

In four experiments tungro disease incidence was recorded along each of the four arms of a cruciform-shaped sampling area. For all varieties used, tungro onset was earlier and disease incidence was greater near the introduced sources than towards the perimeter of the sub-plots or plots.

There were clear disease gradients in the sub-plots and plots with introduced sources and tungro incidence decreased at increasing distances from the source. This is illustrated by the results for two varieties, IR22 and IR26, in the first set of experiments in WS91 and DS92 (Figure 5). In the WS91 experiment, gradients for both varieties in sub-plots with sources were curvilinear and concave in form at 34 DAT. At 48 DAT, the gradient for IR22 had begun to flatten and by 62 DAT flattening was clearly apparent as final disease incidence at the edge of the sub-plots reached almost 50%. By contrast, the gradients for IR26 source plots remained concave even though tungro incidence at the sub-plot borders reached 20%. In plots without sources, tungro incidence increased more rapidly in IR22 than in IR26 but no clear gradients were observed. In the DS92 experiment, when disease levels were considerably lower, the gradients in source plots of each variety and on each date were of similar concave shape and closely resembled those in the early stages of the WS91 experiment (Figure 5). Data for the plots without sources are not shown as disease levels were so low.

Table 2 Final tungro disease incidence (%) in plots with different source disposition in rice variety IR22 in dry (DS93) and wet (WS93) season experiments in 1993

	Source disposition (no. of foci/plot)	Mean final tungro incidence (%)
DS93		
	0	1.2a
	1	4.1b
	25	8.2c
	100	9.8c
WS93		
	0	2.3a
	1	4.1ab
	25	5.4bc
	100	7.0c

Note: In a column in a season, means (average of four replications) followed by a common letter are not significantly different at the 5% level by Duncan's multiple range test. Values were transformed using the arc sine transformation for analysis.

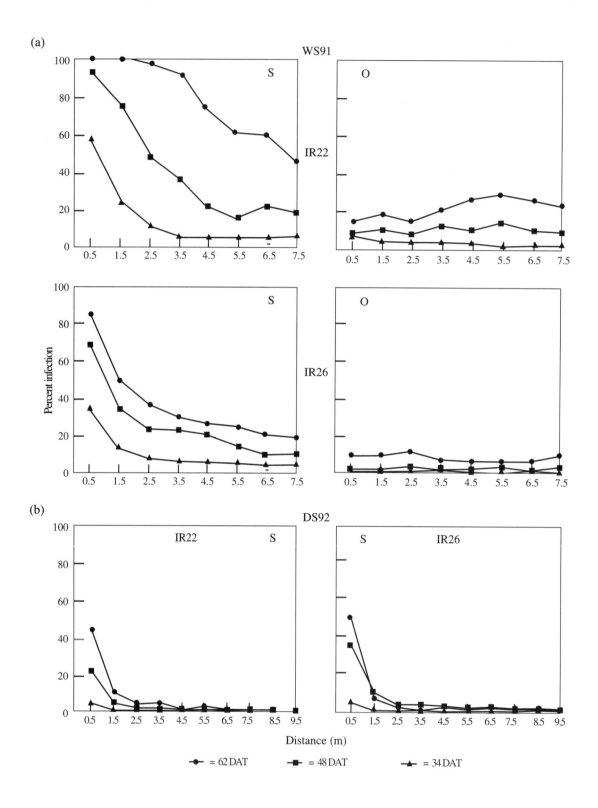

Figure 5   Tungro disease incidence in relation to distance from the centre of the experimental sub-plots or plots in IR22 and IR26 (a) with (S) and without (O) introduced sources of inoculum in the 1991 wet season (WS91) and (b) with introduced sources in 1992 dry season (DS92). DAT = days after transplanting.

Table 3 Average number of leafhopper vectors* collected by sweep net in different varieties in the 1991 wet season (WS) and 1992 dry season (DS) experiments

Variety	Nephotettix virescens				Nephotettix nigropictus				Recilia dorsalis			
	Adults		Nymphs		Adults		Nymphs		Adults		Nymphs	
	WS	DS	WS	DS	WS	DS	WS	DS	WS	DS	WS	DS
IR22	56.5	13.1	139.5	15.1	5.5	1.4	14.0	1.6	1.5	0.3	1.0	0
IR26	79.5	13.6	132.5	11.9	5.0	1.0	16.5	0.9	1.5	0.3	1.0	0.1
IR36	34.5	9.6	67.5	5.2	3.5	1.5	10.0	1.3	1.5	0.4	1.0	0.1
IR72	8.0	6.2	19.0	4.1	1.5	0.8	5.5	1.0	1.0	0.2	0	0.1

*Means of data for plots with and without sources of infection for eight and seven successive weekly catches in the WS and DS trials, respectively.

Leafhopper vector abundance

The most abundant leafhopper vector species in all the experiments was N. virescens, with relatively small numbers of N. nigropictus and R. dorsalis collected. In the first set of experiments in WS91 and DS92, there were large seasonal differences in leafhopper abundance with much greater numbers occurring in the wet season (Table 3). For each variety, leafhopper numbers were similar in plots with and without introduced sources and the data were therefore pooled for analysis. The largest numbers of N. virescens were recorded on IR26 and IR22 and populations were lowest on IR72. Intermediate numbers were found on IR36 in each experiment.

In the two source size experiments (WS92 and DS93), leafhopper numbers were again much higher in the wet season than in the dry season, were greater in IR22 than in IR36 and were similar for all source treatments. By contrast, leafhopper numbers were similar in wet and dry seasons in the source disposition experiments but, as in the previous experiments, there were no differences in abundance between the source treatments.

DISCUSSION

The results from this series of experiments showed that introduced sources of inoculum had an important influence on tungro disease spread. In the DS92 experiment, the final incidence of tungro was less than 1% in each variety where sources were not introduced. Adults and nymphs of leafhopper vector species were recorded as early as 20 DAT (data not shown), but there was little or no indication of disease spread from adjacent plots with sources. The results suggest that few immigrant vectors were infective and that most of the disease spread was secondary, plant-to-plant spread by leafhoppers that acquired tungro viruses from sources within the plots. The effects of introduced sources were more difficult to detect during the WS91 experiment and the WS92 source size experiment when disease levels were higher. This is characteristic of situations where there is also considerable disease spread from outside sources (Thresh, 1976). However, the incidence and distribution of sub-plots or plots without sources were not obviously related to those with. This suggests that inter-plot spread was unimportant and that initial disease spread was mainly from outside the experimental area.

The effect of source size was more clearly demonstrated in the DS93 source size experiment, when even a single-hill introduced source resulted in a significantly greater final disease incidence in a susceptible variety compared with the no-source control. These findings from the source size experiments support those of Shukla and Anjaneyulu (1982) from field cage trials. The results of the source disposition experiments showed that a more dispersed distribution of introduced inoculum sources led to greater disease incidence. Data from other trials conducted on the IRRI farm showed that primary infection due to incoming infective leafhoppers was randomly distributed throughout the plots (Chancellor et al., 1996). Therefore, a dispersed distribution of diseased plants may be expected at the onset of infection in farmers' fields, which would facilitate subsequent disease spread.

The effectiveness of leafhopper resistance in controlling tungro disease was seen in the WS91 and the DS92 experiments. Leafhopper populations on IR72 were small in both seasons and tungro incidence was low, even in the presence of a strong inoculum source. However, the occurrence of populations of N. virescens which have adapted to varieties which were resistant when released highlights the limitations of relying on a single control strategy based on vector resistance (Dahal et al., 1990). The performance of IR26, especially in the WS91 experiment when tungro incidence was high, indicates the value of resistance to RTSV in reducing tungro spread. Hibino et al. (1987) showed that RTSV-resistant varieties become infected with RTBV and develop symptoms if exposed to inoculum from varieties infected with both RTSV and RTBV, but do not become good sources of infection from which further spread of RTBV can occur. The results from our experiments provide further evidence of the effectiveness of RTSV resistance, which is now being incorporated into advanced breeding lines at IRRI. Different sources of resistance to RTSV are being evaluated as an RTSV strain which is virulent to the resistance derived from variety TKM6 has been discovered in one area in the Philippines (Cabauatan et al., 1995).

The large seasonal differences in leafhopper abundance and tungro disease incidence observed in the first two sets of experiments are consistent with other reports that disease levels are generally highest in the wet season (Kondaiah et al., 1976; Rao and Hasanuddin, 1991). Climatic conditions for leafhopper development are generally more favourable in the wet season than in the dry, when less rice is grown. This leads to greater spatial separation between plantings and so the potential for dispersal of vectors and inoculum between fields is less.

The experimental design and the sampling procedures adopted for the first set of experiments proved to be versatile and were changed only slightly for the subsequent experiments. Experience from the WS91 and DS92 experiments resulted in the adoption of slightly larger plots with greater separation distances between plots both within and between blocks to minimize the possibility of inter-plot interference. In retrospect, these changes were not strictly necessary for the dry season experiments when tungro disease levels were much lower but were appropriate for wet season experiments when incidence was generally high. The experience in the WS93 source size experiment, where background levels of disease in the without-source control plots in IR22 reached 40%, showed that even with the bigger plots large exogenous sources of infection located close to the experimental area could lead to too much spread into plots without initial sources. Studies on tungro spread between rice fields in endemic areas, and the major factors influencing such spread, are being conducted elsewhere in the Philippines. The ultimate objective of all these studies is to develop improved control strategies to reduce tungro incidence and to protect the income of rice farmers.

ACKNOWLEDGEMENTS

The authors thank Drs A.G. Cook, H. Koganezawa, J. Holt and K.L. Heong for their help, suggestions and encouragement.

REFERENCES

CABAUATAN, P.Q., CABUNAGAN, R.C. and KOGANEZAWA, H. (1995) Biological variants of rice tungro viruses in the Philippines. *Phytopathology,* **85**: 77–81.

CHANCELLOR, T.C.B., COOK, A.G. and HEONG, K.L. (1996) The within-field dynamics of rice tungro disease in relation to the abundance of its major leafhopper vectors. *Crop Protection,* **15**: 439–449.

DAHAL, G., HIBINO, H., CABUNAGAN, R.C., TIONGCO, E.R., FLORES, Z.M. and AGUIERO, V.M. (1990) Changes in cultivar reactions to tungro due to changes in 'virulence' of the leafhopper vector. *Phytopathology,* **80**: 659–665.

HIBINO, H. and CABUNAGAN, R.C. (1986) Rice tungro-associated viruses and their relations to host plants and vector leafhoppers. pp. 173–181. In: *Tropical Agricultural Research Series* No. 19. Japan: Tropical Agricultural Research Center.

HIBINO, H., TIONGCO, E.R., CABUNAGAN, R.C. and FLORES, Z.M. (1987) Resistance to rice tungro-associated viruses in rice under experimental or natural conditions. *Phytopathology,* **77**: 871–875.

HIBINO, H., DAQUIOAG, R.D., CABAUATAN, P.Q. and DAHAL, G. (1988) Resistance to rice tungro spherical virus in rice. *Plant Disease,* **72**: 843–847.

KONDAIAH, A., RAO, A.V. and SRINIVASAN, T.E. (1976) Factors favouring spread of rice 'tungro' disease under field conditions. *Plant Disease Reporter,* **60**: 803–806.

RAO, P.S. and HASANUDDIN, A. (1991) Incidence of rice tungro virus disease and its vector in South Sulawesi. *Tropical Pest Management,* **37**: 256–258.

SATAPATHY, M.K., CHANCELLOR, T.C.B., TIONGCO, E.R., TENG, P.S. and THRESH, J.M. (1996) The effect of internal and external inoculum sources on tungro disease spread. In: *Rice Tungro Disease Epidemiology and Vector Ecology,* CHANCELLOR, T.C.B., TENG, P.S. and HEONG, K.L. (eds). IRRI Discussion Paper Series No. 19. Los Baños, Philippines: International Rice Research Institute.

SAVARY, S., FABELLAR, N., TIONGCO, E.R. and TENG, P.S. (1993) A characterization of rice tungro epidemics in the Philippines from historical survey data. *Plant Disease,* **77**: 376–382.

SHUKLA, V.D. and ANJANEYULU, A. (1982) Effects of number of leafhoppers and amount and source of virus inoculum on the spread of rice tungro. *Journal of Plant Diseases and Protection,* **89**: 325–331.

SUZUKI, Y., ASTIKA, G.N., WIDRAWAN, K.R., GEDE, G.N., RAGA, N. and SOEROTO (1992) Rice tungro disease transmitted by the green leafhopper: its epidemiology and forecasting technology. *Japanese Agricultural Research Quarterly,* **26**: 98–104.

THRESH, J.M. (1976) Gradients of plant virus diseases. *Annals of Applied Biology,* **82**: 381–406.

THRESH, J.M. (1989) Insect-borne viruses of rice and the Green Revolution. *Tropical Pest Management,* **35**: 264–272.

Paper 4

Simulation modelling of rice tungro virus disease epidemiology

J. HOLT

Natural Resources Institute, The University of Greenwich, Chatham Maritime, Chatham, Kent ME4 4TB, UK

INTRODUCTION

A simulation modelling approach has been applied to the problem of managing rice tungro virus disease (RTVD). The primary aim is to gain a better understanding of how the pathosystem might respond to interventions aimed at disease control. RTVD control options can be divided into two categories: actions to reduce the secondary spread of RTVD within crops already infected, and actions to reduce the risk of primary infection of crops from outside sources.

To investigate options for the control of secondary spread, a spatially structured model was developed which represented a rice plot as a square grid, with each cell of the grid representing a rice hill. RTVD is transmitted by a number of leafhopper species but principally *Nephotettix virescens*. Immigrant vectors were introduced to the grid at random positions, and their movement around the grid determined by a probability distribution of distance moved. With each vector–rice hill contact, virus transmission occurred with a specified probability. The progress of the disease within the plot is the product, therefore, of numerous individual stochastic processes. The details of the model are given in Holt *et al*. (1992a).

A model of similar structure was also developed by Ferris and Berger (1993) to investigate theoretically the implications of different assumptions of virus persistence in the vector. Ferris and Berger did not incorporate vector population processes, whereas it was necessary to do so for our purposes, albeit very simply. Holt and Chancellor (1996) applied the model to the specific question of disease control by removal of diseased plants (roguing). First they calibrated the model for a specific situation using field observations from the Philippines, and then used sensitivity analysis to assess the conditions under which roguing might be effective.

To evaluate the risk of primary infection, the pathosystem was described more simply by modelling populations of vectors and plants rather than individuals. The model is similar to that of Nakasuji *et al*. (1985) for rice dwarf virus. Simplification was possible in the equations which describe infection in the vectors because rice dwarf virus persists in its vector and is transmitted transovarially. The virus also affects vector fecundity and there is a latent period within the vector. Tungro is semi-persistent and none of these complexities applies, but the disease is associated with two viruses: rice tungro spherical virus (RTSV) and rice tungro bacilliform virus (RTBV). For simplicity, however, the model considers them to act as a single infection agent. Five differential equations specify the rates of change of the number of healthy, latently infected and infectious rice plants, and in the number of non-infective and infective vectors. In order to consider movement of inoculum between crops, the equations were repeated for a second location. The impact of vector movement between crops was then examined by including terms in the equations which specified a rate of vector movement from the donor to the recipient crop.

DISEASE SPREAD WITHIN THE CROP

Using the spatially structured model, Holt *et al*. (1992a) investigated the effect on disease progress of the arrival of immigrant vectors. They found that a dual-infective immigrant (i.e. infective with both RTBV and RTSV) contributed 10 to 100 times more to disease progress than did a non-infective vector. Non-infective vectors can, of course, only contribute to disease spread following acquisition of tungro viruses from within the crop itself.

A series of simulations was performed in which the numbers of non-infective and dual-infective immigrants were varied. Simulated disease incidence at the end of the simulation was recorded for different combinations of immigrant numbers. The contour plot of disease incidence (Figure 1) was constructed from the results of this set of simulations. From any point on this contour plot, an increase in the numbers of either non-infective or dual-infective individuals leads to an increase in final incidence. The mix which gives the most rapid increase in incidence, however, is given by a line perpendicular to the tangent of the contour at that point (Figure 1).

22

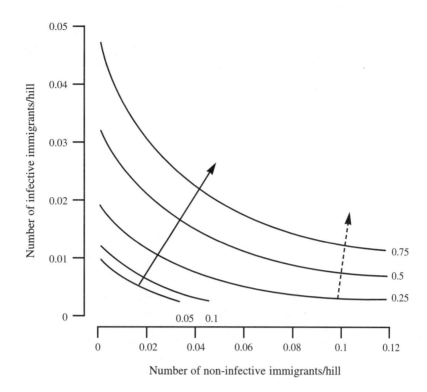

Case a

Case b

Figure 1   Disease incidence simulated by varying numbers of non-infective and dual-infective immigrants. Contour lines indicate final proportion of rice hills infected with RTVD resulting from the arrival of various combinations of infective and non-infective immigrants. The most rapid increase in incidence (arrows) is mainly dependent on inoculum availability at higher vector densities (Case a), and both immigrant number and inoculum availability at lower vector densities (Case b).

Different conclusions are reached, depending on the starting point. When non-infective immigrants are relatively numerous (e.g. >0.1/hill), the 'maximum progress line' is almost parallel to the *y* axis: i.e., infective immigrants are the principal means of disease progress (Case a, Figure 1). The incidence contours are also very close together, indicating that final incidence is highly sensitive to a small input of primary inoculum. This situation may be similar to that encountered in the wet season in southern Luzon island of the Philippines, when vector numbers are relatively high. Significant tungro problems might be expected, therefore, even with relatively low levels of primary inoculum introduced from outside sources.

When numbers of immigrant vectors are low, then the 'maximum progress line' reflects a constraining influence of vector numbers (Case b, Figure 1). An important increase in disease can occur by addition of more non-infective immigrants alone. In such a situation, therefore, tungro incidence might be expected to show a stronger relationship with vector number than in the 'wet-season case' discussed above. Low vector numbers are a characteristic of the dry-season crop in southern Luzon, when it is unusual for serious tungro problems to occur. In the 'dry-season case' the contours are much further apart indicating that much more substantial amounts of primary inoculum (i.e. infective immigrants) are required to give a high final incidence.

CONTROLLING DISEASE SPREAD

The spatial grid model was also used to gain some indication of the likely impact of two possible tactics to control disease spread: roguing and vector control with insecticides. The efficacy of both these methods is likely to depend on the extent of the prevailing infection pressure. In each case, therefore, a series of simulations was performed in which the infection pressure was varied by changing the number of infective immigrants entering the crop.

For the simulated roguing, it was assumed that any diseased plants were removed every four days for the first eight weeks of the crop. This is probably too frequent and too demanding to be feasible in most cases but gives some indication of the potential benefits. For the particular case represented by this series of simulations, roguing was successful in curtailing disease spread up to moderate levels of inoculum pressure. However, the number of plants which had to be removed was prohibitively high at all but the lowest rates of infective vector immigration (Figure 2a). Holt and Chancellor (1996) give a more complete analysis of the roguing issue.

Vector control with insecticides was also simulated. For example, it was assumed that 95% of all leafhoppers within the crop were killed 20 days after transplanting (DAT). Individual spray applications are unlikely to be so efficient in practice but more than one application could be applied to achieve greater mortality. There was also assumed to be no subsequent increase in leafhopper population growth rate due to mortality caused to its natural enemies. Only the case of a single application at 20 DAT was examined here. At low infection pressure, a reasonable reduction in final disease incidence over the no-spray case is achieved: about 60% at the lowest immigration rate (Figure 2b). The gains compared to the no-spray case decline rapidly with increasing infection pressure, however. In most cases, therefore, it is likely that insecticide sprays would be of rather limited efficacy. They may also be inadvisable due to the resurgence risks associated with early sprays (Holt *et al.*, 1992b).

RISK OF PRIMARY INFECTION

The basic model for a single location, used to investigate primary infection risk from outside sources, is summarized in Figure 3. The numbers of healthy, latently infected and infectious rice hills are denoted by X, Y and Z, respectively; and non-infective and infective vectors are denoted by U and W, respectively. The latent period between acquisition by the vector and the ability to transmit was assumed to be negligible. The numbers of rice hills in each class change at the following rates:

$$\frac{dX}{dt} = -\frac{pWX}{t} \tag{1}$$

$$\frac{dY}{dt} = \frac{pWX}{t} - rY \tag{2}$$

$$\frac{dZ}{dt} = rY \tag{3}$$

where p = inoculation rate
t = age of the rice crop
r = reciprocal of mean latent period (in the rice)

The numbers of vectors in each class change at the following rates:

$$\frac{dU}{dt} = aN\left(1 - \frac{N}{K}\right)\left(1 - \frac{t}{b}\right) + iW - cU - qZU \tag{4}$$

$$\frac{dW}{dt} = qZU - (c + i)W \tag{5}$$

where a = birth rate of vectors
K = maximum number of vectors per plant
b = crop age at which vector birth rate declines to zero
i = reciprocal of mean retention period of virus in vector
c = death rate of vectors
q = acquisition rate
N = total vector number, $U + W$

(a)

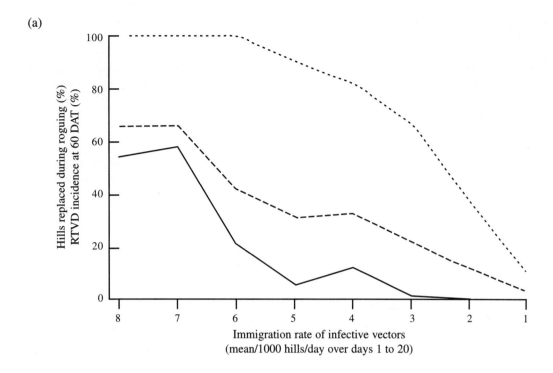

(b)

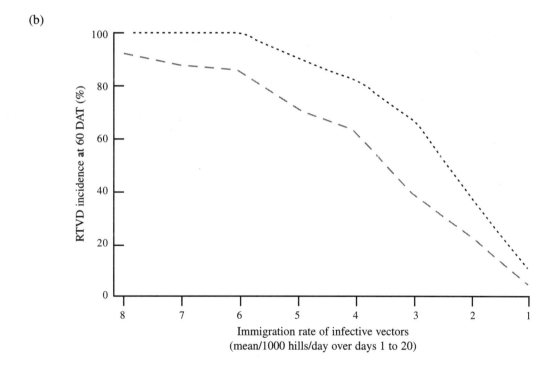

Figure 2 Simulated RTVD incidence at 60 DAT resulting from different levels of inoculum pressure, i.e. number of infective immigrant vectors.
(a) ———— with roguing every four days for the first eight weeks of the crop; . . . . . . with no control measures; - - - - - indicates the number of hills requiring replacement in the simulated roguing treatment.
(b) – – – – – with a single insecticide application causing 95% mortality in the leafhopper population at 20 DAT; - - - - - with no control measures.

25

As can be seen from Equations 1 to 3, the initially healthy plant population (X) gradually becomes latently infected at a rate dependent upon the number of infective vectors (W) and the inoculation rate (p). Latent infections become infectious at a rate (r): $1/r$ is the mean latent period. Virus inoculation rate is assumed to be inversely proportional to plant age. There is evidence to suggest an effect of plant age, but the exact form of the relationship is unknown. The simplest possible assumption was used that the rate is inversely proportional to t. This is analogous to the log-logistic model of disease progress (Jeger, 1983).

Vector birth rate (a) is assumed to have a logistic form constrained to a maximum carrying capacity (K). Birth rate also declines with plant age and, as with the inoculation rate discussed above, a linear decline with increasing plant age is assumed. The population of healthy vectors becomes infective at a rate dependent on the number of infectious plants and the acquisition rate (Equation 4). Once infective, vectors are assumed to retain infectivity for three days ($1/i$ = mean retention period). Infective vectors revert therefore to being non-infective at a rate i (Equation 5). Vector death rate is assumed to be constant and unaffected by the presence of virus.

To simulate the movement of individuals from one crop to another, the set of five equations was repeated for a second location. A number of different assumptions about vector movement was investigated; the simplest was that the rate of vector arrival at the recipient crop was inversely proportional to the distance between the two crops. This assumption implies that, within the limits investigated, there is an equal probability of vectors travelling any distance so that the reduction in numbers as distance increases is due simply to radial dilution (i.e. a linear increase in the length of the front over which vectors are distributed as distance travelled increases). To do this, the terms mU and mW were deducted from Equations 4 and 5 for the donor location and the terms mU/s and mW/s added to the same equations for the recipient location, respectively, where m = rate of emigration from the donor location and s = distance between locations.

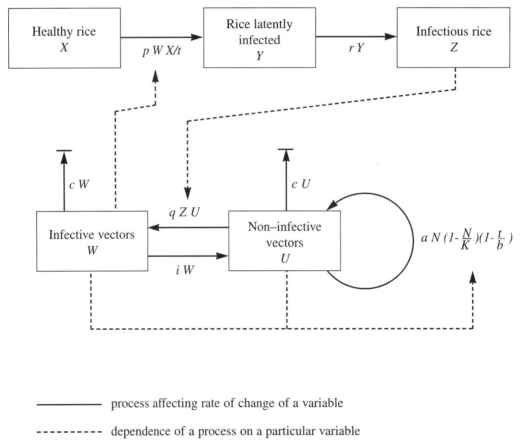

—————— process affecting rate of change of a variable

---------- dependence of a process on a particular variable

Figure 3   Five-variable differential equation model for a single location.

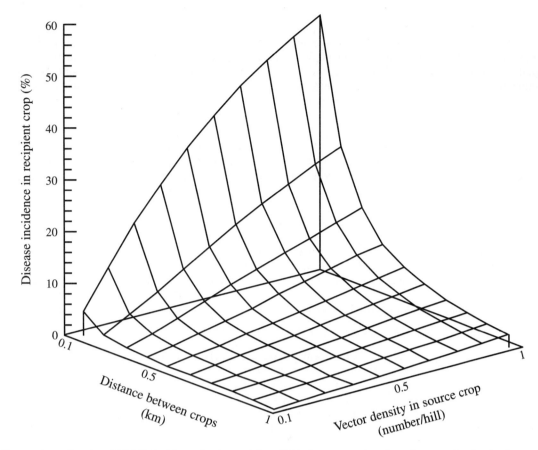

Figure 4 Simulated RTVD incidence in the recipient field shown as a function of vector density in the source and the distance between the two fields. Results from the five-equation model with source and recipient fields linked by movement of vectors.

SIMULATION ANALYSIS OF RTVD INCIDENCE IN RELATION TO DISTANCE FROM SOURCE

Initial estimates of parameter values were: $p = 0.01$, $r = 0.14$, $a = 0.14$, $K = 30$ $(X + Y + Z)$ (i.e. 30/hill), $b = 80$, $I - 0.33$, $c = 0.05$, $q = 0.1$, $m = 0.05$. It is intended to parametrize the model more precisely using NRI/IRRI (Natural Resources Institute/International Rice Research Institute) tungro project field data and information from the literature. Simulations were performed using 4th order Runge-Kutta numerical integration (Press *et al.*, 1989).

A series of simulations was performed for a range of distances between crops. The vector population size in the donor crop was also varied and the final disease incidence in the recipient crop was recorded in each case. Figure 4 shows how steeply final disease incidence declined with distance from the donor crop. Clearly, limited weight can be attached to the absolute values without more precise parametrization and sensitivity analysis. The suggestion from these initial simulation experiments, however, is that the zone of risk is likely to extend to several hundreds of metres rather than several kilometres.

The simplistic assumption made here, that flight distances follow a uniform distribution, is incorrect. In fact, the results of flight mill experiments (Cooter and Winder, 1994) suggest that there is an extremely skewed distribution and most individuals fly for only limited periods implying very short distances. Some simulations were also performed assuming an exponential decline in insect settling with increasing distance from the source. As to be expected, the incidence in the recipient crop was further reduced for a given separation distance.

DISCUSSION

Two models have been used to help our thinking in relation to the problems of managing RTVD. The first model is described in detail elsewhere (Holt *et al.*, 1992a; Holt and Chancellor, 1996). Here some results are presented of simulations to investigate two possible measures that have been advocated to prevent disease spread: roguing diseased plants and use of insecticides to kill the vector. Once inoculum pressure in a field reaches moderate levels, the results of the models suggested that it is extremely difficult to curtail the spread of the disease by more than a small amount. In the case of roguing, results of field trials (E.R. Tiongco, personal communication) also lead to the conclusion that roguing operations can only keep pace with the advance of the disease when the amount of primary inoculum or the subsequent rate of spread is low.

Again, for simulated insecticide use, the greatest reduction in incidence compared with the untreated was achieved when infection pressure was lowest. It is now generally recognized that insecticide use on rice early in the growing season carries a very high risk of inducing pest resurgence. The current IRRI recommendation is not to spray during the first 40 days of the rice crop. Even if insecticides show promise of greater efficacy, there would, therefore, be reluctance to recommend their use.

The situation with regard to influencing the risk of primary infection is perhaps less pessimistic. The initial explorations with the five-equation model described here suggest that the risk of significant tungro incidence in a crop declines steeply with distance from the source of inoculum. The steepness of the decline depends upon assumptions about vector movement. If it is assumed that vectors emigrate along random trajectories (or along random trajectories within a segment of possible trajectories, rather than a complete circle), then the effects of radial dilution alone are sufficient to have a major impact. Declining flight probability with distance is an additional effect that reinforces this trend.

Although the model has not yet been parametrized accurately, the conclusions concerning risk in relation to distance are consistent with a more recent field survey of tungro incidence spanning several hundred fields over the period of a year. Distances between existing and newly infected fields were measured from the survey data. Newly infected fields were seldom more than 400 m from the nearest one already infected. This contrasted with 'vulnerable' but not infected fields which were distributed over a greater range of distances within the 1.6 × 0.7 km survey site (Chancellor *et al.*, 1995; Holt *et al.*, 1996).

There are important implications for tungro avoidance. Whilst ideally it would be prudent to avoid planting within, say, 500 m of an existing RTVD-infected field, such flexibility is seldom likely to be practicable. More appropriate may be a strategy for the deployment of RTVD-resistant varieties. It may be necessary to grow varieties with RTVD resistance only within quite restricted locations where the disease is locally serious. Such a strategy of targeted deployment of resistant material is likely to increase the durability of resistance. Repeated evidence of resistance breakdown is already apparent in a sequence of vector-resistant varieties introduced since the 1960s. Durability remains an important issue both for the deployment of virus-resistant varieties from conventional breeding programmes and for resistant material based on transgenic lines (Khush, 1994).

REFERENCES

CHANCELLOR, T.C.B., TIONGCO, E.R. and HOLT, J. (1995) Spread of rice tungro virus disease between fields. pp. 27–28. In: *Sixth International Plant Virus Epidemiology Symposium, Jerusalem, 23–29 April 1995*. Bet Dagan, Israel: Phytopathological Society of Israel.

COOTER, R.J. and WINDER, D. (1994) Flight performance of leafhopper vectors of rice tungro disease. *Proceedings of the Third International Conference on Tropical Entomology, ICIPE, Nairobi, 30 October–4 November 1994*. Nairobi: International Centre for Insect Physiology and Ecology. [Abstract]

FERRIS, R.S. and BERGER, P.H. (1993) A stochastic simulation model of epidemics of arthropod-vectored plant viruses. *Phytopathology,* **83**: 1269–1278.

HOLT, J. and CHANCELLOR, T.C.B. (1996) Simulation modelling of the spread of rice tungro virus disease: the potential for management by roguing. *Journal of Applied Ecology,* **33**: 927–936.

HOLT, J., CHANCELLOR, T.C.B and SATAPATHY, M.K. (1992a) A prototype simulation model to explore options for the management of rice tungro virus disease. pp. 973–980. In: *Proceedings of Brighton Crop Protection Conference: Pests and Diseases, Brighton, November 1992*. Farnham, Surrey, UK: British Crop Protection Council.

HOLT, J., WAREING, D.R. and NORTON, G.A. (1992b) Strategies of insecticide use to avoid resurgence of *Nilaparvata lugens* in tropical rice: a simulation analysis. *Journal of Economic Entomology*, **85**: 1979–1989.

HOLT, J., CHANCELLOR, T.C.B., REYNOLDS, X. and TIONGCO, E.R. (1996) Risk assessment for rice planthopper and tungro disease outbreaks. *Crop Protection*, **15**: 359–368.

JEGER, M.J. (1983) Analysing epidemics in time and space. *Plant Pathology*, **32**: 5–11.

KHUSH, G. (1994). Breaking the yield frontier of rice. pp. 17–21. In: *Proceedings of the ODA/IRRI/ BBSRC Meeting, Food Security in Asia, Royal Society, London, 1 November 1994*. London: Royal Society.

NAKASUJI, F., MIYAI, S., KAWAMOTO, H. and KIRITANI, K. (1985) Mathematical epidemiology of rice dwarf virus transmitted by green leafhoppers: a differential equation model. *Journal of Applied Ecology*, **22**: 839–847.

PRESS, W.H., FLANNERY, B.P., TEUKOLSKY, S.A. and VETTERLING, W.T. (1989) *Numerical Recipes in Pascal*. Cambridge: Cambridge University Press.

Paper 5

Epidemiology-oriented forecasting of rice tungro virus disease in asynchronous rice cropping areas

Y. SUZUKI[1]*, G.N. ASTIKA[2], K.R. WIDRAWAN[2], G.N. GEDE[2], N.S. ASTIKA[2], N. SUWELA[2], G.N. ARYAWAN[2] and SOEROTO[1]
[1]*Directorate of Food Crop Protection, Pasarminggu, Jakarta, Indonesia*
[2]*Crop Protection Center VII, Denpasar, Bali, Indonesia*

INTRODUCTION

The implementation of large-scale synchronous rice cropping with a definite fallow period between cropping seasons has been remarkably successful in controlling rice tungro virus disease (RTVD), commonly referred to as tungro or rice tungro disease. This cultural control method has been quite effective in eradicating virus sources and in suppressing the immigrant population density of the main leafhopper vector, the green leafhopper (GLH), *Nephotettix virescens*, which is monophagous on rice (Cabunagan and Hibino, 1989; Loevinsohn and Alviola, 1991; Sama *et al.*, 1991; Taib, 1987). However, further implementation of this approach is hindered by various socio-economic and socio-cultural constraints in many tungro-endemic areas. In addition, synchronous cropping may lead to increased problems due to the brown planthopper, *Nilaparvata lugens* (Sawada *et al.*, 1993).

Control measures so far used in asynchronous cropping areas include cultivation of vector-resistant or tungro virus-resistant varieties, eradication (roguing) of infected plants and insecticide application. The effectiveness of eradicating infected plants and insecticide applications is limited (Estano and Shepard, 1989; Raga *et al.*, personal communication) because immigration of infective vectors takes place continuously in asynchronous cropping areas. Moreover, infected plants become sources of viruses before they develop visible symptoms (Hino *et al.*, 1974) and frequent roguing of infected plants is very laborious. Field experiments made under high immigration pressure of infective GLH showed that both roguing of infected plants and insecticide application were unsuccessful in significantly reducing tungro incidence if these measures were adopted piecemeal by individual farmers operating independently (Suzuki *et al.*, unpublished).

Cultivation of resistant varieties has been the sole feasible and effective practice in controlling tungro in asynchronous cropping areas. Yet there has been a vicious cycle of intensive cultivation of resistant varieties followed by the development of virulent biotypes against them. It becomes increasingly difficult to breed varieties that fulfil the requirements of GLH resistance together with high yield and high quality. Resistant varieties should ideally be used only when necessary to avoid rapid breakdown of resistance. Even in tungro-endemic areas, outbreaks of tungro have not occurred frequently. Development of forecasting technology is therefore essential for the proper use of tungro-resistant varieties.

This paper presents some results of epidemiological studies on tungro disease in Bali, Indonesia, under the Indonesia–Japan Joint Programme on Food Crop Protection implemented by the Directorate of Food Crop Protection, Ministry of Agriculture, Indonesia and the Japan International Co-operation Agency.

STUDY AREAS AND METHODS

Since the first outbreak of tungro was recorded in Bali in the wet season of 1980/81, the disease has been endemic in the three regencies of Badung, Tabanan and Gianyar. Rice has been cultivated on an average of c. 120 000 ha/year in the three regencies in the past 10 years; large-scale synchronous cropping has never been implemented in Bali. Irrigation in this province has been managed by farmers' associations (*subaks*), each of which consists of 50–200 farmers. Rice is cultivated either synchronously or asynchronously within each *subak*.

To compare the population dynamics of GLH and tungro occurrence between small-scale synchronous (i.e. within a *subak*) and asynchronous cropping areas in Bali, a total of eight fields were studied in Badung and Gianyar in the dry season of 1987 and the wet season of 1987/88. Seedlings of the GLH-susceptible varieties Krueng Aceh or IR36 were transplanted 21 days after sowing in farmers' fields using about three seedlings/hill at 25 × 25 cm spacing. Insecticides were not applied and in other respects the maintenance of the fields was standard. GLH adults and nymphs were collected weekly by 75 sweeps of a sweep net covering c. 300 hills. All the adults were dissected under a binocular microscope to check for

---

*Present address: Kyushu National Agricultural Experiment Station, Nishigoshi, Kumamoto 861-11, Japan.

parasitism. GLH egg density was estimated by dissecting plants in 1580 hills, sampled randomly. Egg mass samples were kept individually in small glass vials with a piece of wet cotton, and inspected for parasitism 10 days later. Additional population assessments of GLH and natural enemies were made at 4, 8 and 12 weeks after transplanting (WAT) with a FARMCOP suction sampler (Cariño et al., 1979) at 20 four-hill locations. For further details of the census methods see Aryawan et al. (1993a).

Intensive studies of tungro spread and GLH population dynamics in paddy fields were made at Padang Galak in Badung where lowland asynchronous rice planting has been practised for decades. All growth stages of rice and many tungro-infected fields occurred throughout the study period. Plantings for routine population sampling were made in farmers' fields at intervals of about one month from July 1988 to March 1990. Tungro-infected hills were mapped weekly based on visual diagnosis in 10×10 m plots containing 1600 hills. Census methods for GLH and natural enemies, and identification of GLH generations are described in Aryawan et al. (1993b) and Suzuki et al. (1992).

At weekly intervals from 2–10 WAT, 10 hills that developed visible symptoms during the previous week were labelled and harvested individually at maturity to estimate yield. Ten hills that did not develop tungro symptoms until harvest were sampled randomly in the same plot to act as control hills. Yield loss assessment was based on the dry seed weight.

Data on monthly rainfall and the tungro-infected areas in each regency were obtained from pest observers' reports. The Bali Provincial Agriculture Service provided relevant statistics for analysis.

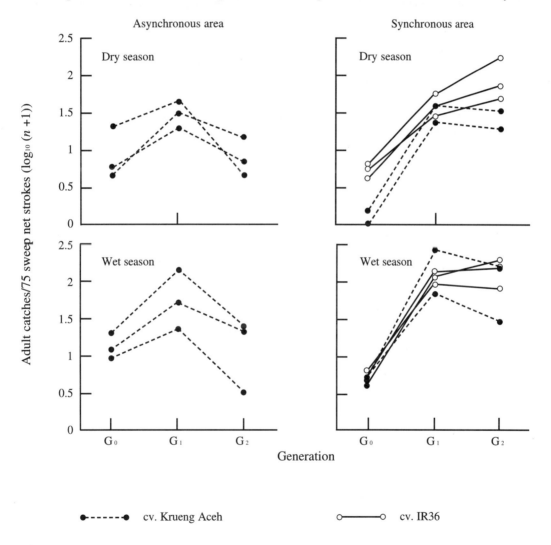

Figure 1   Population growth patterns of two cultivars of *N. virescens* adults in asynchronous (three paddy fields) and small-scale synchronous (five paddy fields) rice cropping areas during the dry season 1987 and wet season 1987/88 (after Aryawan et al., 1993a).

POPULATION DYNAMICS OF GLH

Green leafhopper population growth patterns in asynchronous cropping areas were quite different from those in small-scale synchronous cropping areas (Figure 1). In asynchronous areas, the adult density of GLH on Krueng Aceh reached a maximum in the first generation (G_1) and then greatly decreased in the second generation (G_2), irrespective of the season. In contrast, population growth rates were much higher and population densities of G_1 and G_2 were similar in small-scale synchronous cropping areas.

Both natural enemy abundance and meteorological factors failed to explain the difference between the two cropping regimes (Aryawan et al., 1993a). Widiarta et al. (1990) compared the population growth pattern of GLH between an asynchronous and a large-scale synchronous rice planting area and also found very low population growth rates of G_2/G_1 to be characteristic of the asynchronous area. They suggested that this may be due to the differential effect of inter-field adult dispersal. Emigration and immigration in individual paddy fields tend to compensate for each other in synchronous areas, whereas in asynchronous areas the G_1 adult density decreases due to emigration to young rice. This effect of adult dispersal may also account for the difference between the small-scale synchronous and asynchronous cropping areas.

Using FARMCOP sampling on 17 paddy fields at Padang Galak, Aryawan et al. (1993b) showed that GLH population density was highest in the immigrant generation (G_0) or G_1 and lowest in G_2 in asynchronous areas. (Figure 2 shows changes in the mean adult density with rice stage.) Although the adult density was highest in G_1 (5–8 WAT), the mean percentage of females with mature ovaries was much less in G_1 than in G_0, resulting in very low G_2 egg densities (Figure 3). This is circumstantial evidence for high emigration rates of G_1 adults in asynchronous cropping areas.

Bali farmers in asynchronous cropping areas usually cultivate rice successively with no definite break between crops, starting land preparation within four weeks of harvest. The mean GLH adult density on stubble and ratoon rice up to four weeks after harvest was much lower that that recorded earlier in the growing crop (see Figure 2).

TUNGRO INCIDENCE

Tungro disease occurred in all 17 census plots in Padang Galak (Figure 4). Severe infection (>90% hills infected) occurred every season, though the infection tended to be most serious in fields transplanted in the late wet season (January–March). Cumulative percentages of infected hills at 2 WAT were low in all the plots, indicating little infection in seed-beds. This can be attributed to very low GLH density in seed-beds in the asynchronous cropping area; the number of GLH adults and nymphs collected by a sweep net (40 cm in diameter) in the 17 seed-beds one day before transplanting was only 3.1±2.5/10 sweeps.

Peak infection occurred on average at 7 WAT (Figure 5). Assuming an incubation period of c. two weeks for tungro in rice plants (Astika et al., 1992), it was concluded that peak tungro transmission occurred around 5 WAT by G_1 nymphs of GLH (Figure 5). A similar result was obtained by Suzuki et al. (1989) based on the field study made at three asynchronous and five small-scale synchronous cropping areas in Bali. While the cumulative percentage of infected hills increased monotonously as rice grew older, GLH population density dropped sharply in G_2. The acquisition and transmission rates of tungro viruses by GLH also decrease as plants age (Narayanasamy, 1972; Suwela et al., 1992); this explains why G_2 vectors seldom play a significant role in spread of tungro within fields.

YIELD LOSS

Hills infected by G_0 adults are mostly distributed singly in paddy and later a more clumped or patchy distribution of diseased hills appears as a result of secondary tungro transmission by G_1 nymphs. The mean yield loss sustained by an infected hill depended on the stage disease occurred and the location of the hill (Table 1). Diseased hills located within a tungro patch suffered higher yield loss than isolated diseased hills or those at the periphery of a tungro patch, irrespective of the disease stage. This is because almost all the plants in a hill within a tungro patch were diseased, due to secondary transmission by G_1 nymphs. In contrast, isolated and peripheral diseased hills often contained healthy plants which partially compensated for their diseased neighbours even if disease occurred early.

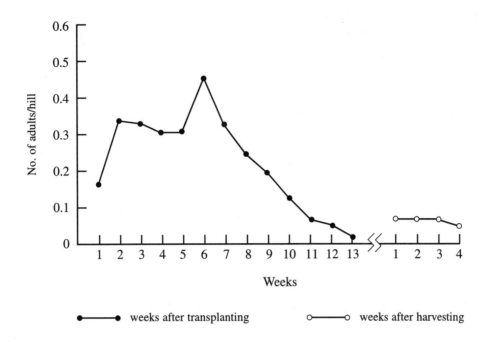

Figure 2   Phenology of populations of *N. virescens* adults in 17 paddy fields transplanted between July 1988 and March 1990 and in stubble after harvest at Padang Galak, an asynchronous rice cropping area in Bali.

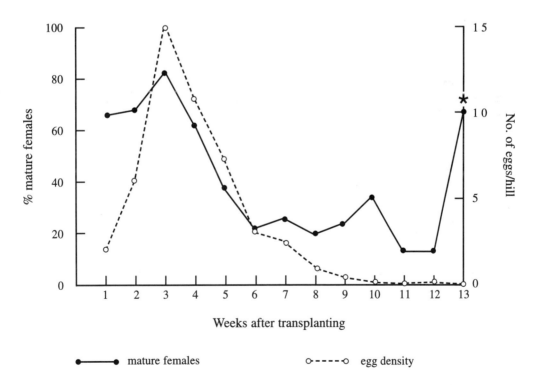

Figure 3   Phenology of the mean percentage of female *N. virescens* with mature ovaries and the mean egg density in 17 paddy fields transplanted from July 1988 to March 1990 at Padang Galak. The point with an asterisk is based on small sample size (total number of females = 6).

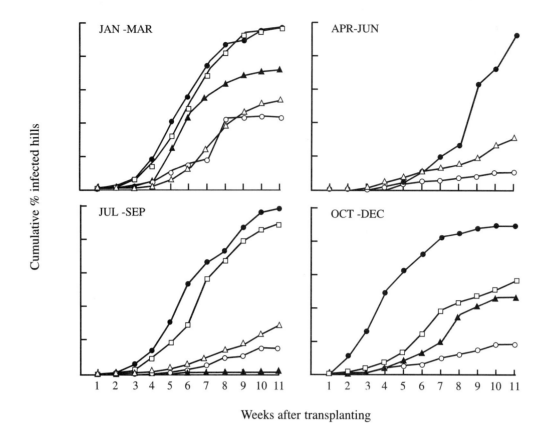

Figure 4   Incidence of tungro disease in 17 paddy fields transplanted in four different periods during 1988. Symbols indicate individual paddy fields.

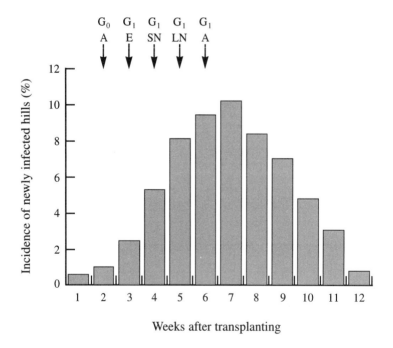

Figure 5   Weekly changes in incidence of newly diseased hills among healthy hills in previous week in 17 paddy fields at Padang Galak. Arrows indicate occurrence of population peaks of different stages of *N. virescens* G_0 or G_1 generation. Adults (A), eggs (E), 1st–3rd instar nymphs (SN) and 4th–5th instar nymphs (LN).

Table 1 Effect of diseased stage and position on percentage yield loss sustained by RTV-diseased hills compared with healthy controls

Diseased stage (WAT)	Location of hill in relation to tungro patch		
	Inside	Periphery	Outside
2–4	92.3 ± 4.5a	63.4 ± 6.0b	45.7 ± 19.7b
5–7	78.2 ± 5.3a	58.2 ± 15.2b	45.8 ± 10.4c
8–9	71.6 ± 10.2a	51.6 ± 14.1b	32.5 ± 17.6c

Note: In a row, means followed by the same letter are not significantly different at the 5% level (Tukey's multi-range test).

These results indicate that G_1 nymphs play the most important role in causing tungro infections at a stage when the greatest yield reduction of rice is caused.

PREDICTION OF TUNGRO OCCURRENCE

In transmission tests with GLH caught in tungro-infected paddy fields, Suwela *et al.* (1992) established that the percentage of infective GLH is proportional to the percentage of infected hills where incidence of tungro is low. From this relationship the infective GLH density in young paddy fields can be estimated. The cumulative percentage of infected hills at harvest is predictable with a high degree of accuracy by an index of infective G_1 nymphal density and infective vector index (IVI), expressed as density of G_1 adults and 4th–5th instar nymphs × percentage of diseased hills at 6 WAT (Suzuki *et al.*, 1992). However, for practical purposes, tungro incidence should be predicted much earlier and before secondary within-field spread occurs.

The G_1 healthy egg density was almost proportional to G_0 adult density (Figure 6). The main cause of the considerable fluctuation in survival rate of G_1 nymph and adults has been considered to be the fluctuation in the emigration rate of G_1 adults and not nymphal survival (Aryawan *et al.*, 1993b). Hence the relative abundance of G_1 nymphs may be well represented by G_0 adult density. The percentage of diseased hills at any time before 6 WAT may serve as an appropriate measure of total infection caused by G_0 adults. It follows that the IVI — even before 6 WAT — is helpful in predicting tungro incidence at harvest as shown in Figure 7. However, accurate assessment of GLH populations is often difficult for farmers especially when the population density is low. Control thresholds based solely on the percentage infected hills showing visible symptoms were therefore established for the measurements made at 2–5 WAT (Suzuki *et al.*, 1992). Since the control thresholds are established empirically, they are not applicable where GLH density is very high.

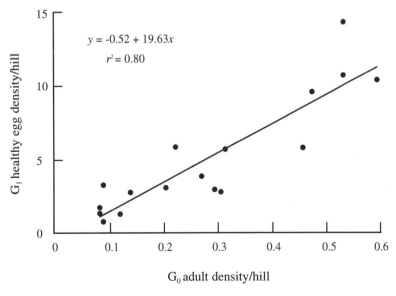

Figure 6 Relationship between adult density of immigrant generation and healthy egg density of first generation in *N. virescens* at Padang Galak.

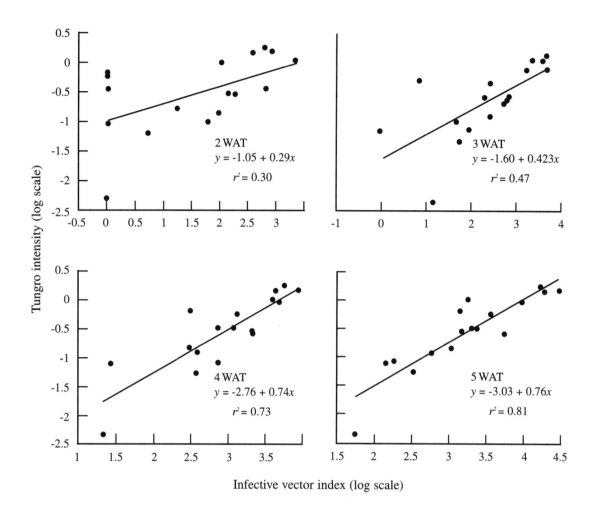

Figure 7 Regression of the tungro intensity at 10 WAT on infective vector index (IVI) at 2–5 weeks after transplanting. Tungro intensity = log(no. healthy hills/total no. hills) at 10 WAT; IVI = (no. *N. virescens* adults/80 hills) × (% diseased hills with obvious tungro symptoms).

TUNGRO FORECASTING AT REGENCY LEVEL

Monthly fluctuations in the incidence of newly infected areas and rainfall in Gianyar, which is one of the most serious tungro-endemic regencies, are shown in Figure 8 for 1985–90. GLH-susceptible varieties such as Krueng Aceh, IR36 and IR64 were cultivated in more than 80% of paddy fields every season; IR36 and IR64 were released in Bali as being GLH-resistant, although they behaved as susceptible by 1980 and immediately after release, respectively. The newly infected area was usually lowest in late dry season (July–September) and increased around November corresponding to the start of the wet season (monthly rainfall ≥200 mm). After attaining a peak around December, it dropped and increased again late in the wet season or early in the dry season. These bimodal fluctuations were also observed in the other tungro-endemic regencies (Suzuki *et al.*, 1992).

The dependence of the newly infected area at month t on 17 variables, rainfall (precipitation P) at $t \sim t\text{-}5$, newly transplanted area (NT) at $t \sim t\text{-}5$ and newly infected area (NI) at $t\text{-}1 \sim t\text{-}5$ was analysed with the six years of data. The simple regressions are highly significant ($p<0.01$) for 6 of the 17 variables (Table 2). Since reports by pest observers on the rice stage of newly infected fields were within a range of 4 and 7 WAT with the mode at 6 WAT, immigration of GLH to the newly infected fields may have occurred, usually *c.* one month before the report was made. The negative regression coefficient of $NT_{t\text{-}1}$ and the positive regression coefficients of $NT_{t\text{-}3}$ and $NT_{t\text{-}4}$ suggest that transplanting is increasingly risky under conditions where more of the rice is at a late stage. This suggests that tungro occurs with a high probability if transplanting is done when there is much immigration of GLH. The other variables which are highly significant include $P_{t\text{-}1}$ and $NI_{t\text{-}1}$. It is not clear why NI_t is related negatively to $P_{t\text{-}1}$ and positively to P_t. This agrees with the observation that tungro increases at the beginning of the wet season, but is inconsistent with the increase in tungro during the transition period from the wet to the dry season. The importance of $NI_{t\text{-}1}$ is discussed below.

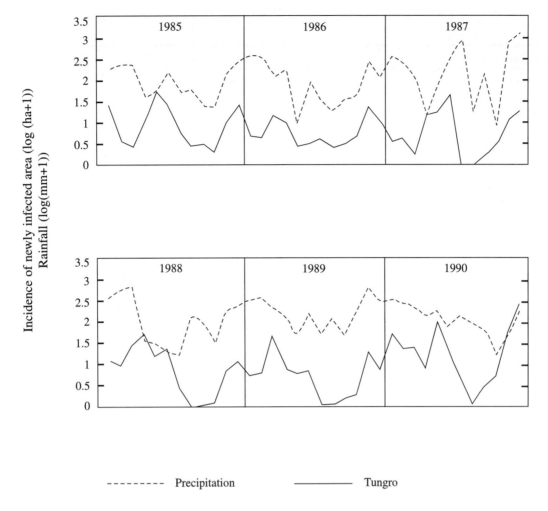

- - - - - - - Precipitation ——————— Tungro

Figure 8 Monthly fluctuations in the incidence of newly infected area [log (ha+1)] and rainfall [log (mm+1)] in Gianyar regency, Bali, 1985–90.

As some of the variables created as independent in the simple regression analysis are interrelated, stepwise regression analysis was employed; it showed that 48% of the variance in the monthly incidence of new infection is explained by three variables (Table 3). Abundance of NI_{t-1} and NT_{t-3} are two major factors responsible for the monthly new infection. Because NI_{t-1} is the tungro-infected area at about 6 WAT at t-1, it is regarded as the most important source of infective migrants which invade and cause initial infection. The proportion of infective GLH in tungro-infected fields at 5–7 WAT is known to be very high (Suwela *et al.*, 1992). By contrast, NT_{t-3} is considered to be the most important source of GLH migrants. NT_{t-3} is at about 8–13 WAT at t-1. This rice stage is the main source of migrants in small-scale synchronous areas which are more extensive than the asynchronous areas in Gianyar.

It is notable that the newly infected area (NI_t) does not depend on the tungro-infected area at t-2 (NI_{t-2}) whose rice stage is similar to the major source of GLH migrants (NT_{t-3}). Why does the major source of infective migrants differ from that of total migrants? A possible explanation is that GLH adults emigrate from tungro-infected paddy fields earlier than those in tungro-free fields. That the emigration rate of G_1 adults is high in fields of high tungro intensity was found in the detailed studies made at Padang Galak (Suzuki *et al.*, unpublished).

Tungro occurrence is widely believed to depend on season. The above preliminary analysis suggests that rainfall plays a minor role in determining the magnitude of tungro infection, though the effect is significant. It is likely, however, that rainfall affects tungro transmission directly and indirectly through many processes, and that its overall effect on tungro occurrence is complex. Further study is needed to analyse the effect of rainfall and other meteorological factors.

Suzuki *et al.* (1993) provided other empirical rules for forecasting tungro occurrence at the regency level.

Table 2 Regression of newly infected area/month on 17 variables in Gianyar Regency, Bali, 1985

Independent variable	Std. coefficient	F
Precipitation t	0.268	5.440*
Precipitation t-1	−0.388	12.437***
Precipitation t-2	−0.154	1.708
Precipitation t-3	0.030	0.065
Precipitation t-4	0.149	1.583
Precipitation t-5	0.187	2.539
New transplanting t	−0.333	8.703**
New transplanting t-1	−0.413	14.353***
New transplanting t-2	−0.020	0.029
New transplanting t-3	0.463	19.083***
New transplanting t-4	0.522	26.222***
New transplanting t-5	0.124	1.093
New infection t-1	0.483	21.250***
New infection t-2	0.034	0.080
New infection t-3	−0.252	4.729*
New infection t-4	−0.265	5.271*
New infection t-5	−0.185	2.487

Note: Log-transformed values were used for amount of precipitation and newly transplanted area.
* = $p<0.05$; ** = $p<0.01$; *** = $p<0.001$

Table 3 Summary of stepwise regression analysis of monthly new infection in Gianyar Regency, Bali, 1985

Variables in equation	Coefficient	Std. coefficient	F
New infection t-1	0.000149	0.473	28.922*
New transplanting t-3 (log)	0.439	0.422	21.753***
Precipitation t (log)	0.217	0.205	5.094*
Correlation coefficient (R)	0.691		
R-squared	0.478		
Intercept	−0.343		

* = $p<0.05$; ** = $p<0.001$

DISCUSSION

Field studies in asynchronous rice cropping areas in Bali indicate that G_1 nymphs of GLH are mainly responsible for both tungro spread and tungro-induced yield reduction in individual paddy fields. Although the cumulative percentage of infected hills did not exceed 80% until after 6 WAT in the 17 fields (see Figure 4), it was not unusual to observe that more than 80% of hills were infected in some young paddy fields at 4 WAT in Bali. In these fields there was a marked difference in the percentages of infected hills at the margins of a field and within: <20% at the margins compared with >80% within (Suzuki et al., unpublished). This large difference in tungro incidence indicates that the main spread of tungro occurred after transplanting. Seed-bed infection alone is seldom considered sufficient to cause serious loss of yield, unless the nursery period is so unusually prolonged as to facilitate secondary infection before transplanting.

GLH migrants which spread tungro between fields are G_1 adults produced mainly in paddy fields 5–8 WAT in asynchronous cropping areas. Production of GLH migrants in such young crops and high virus acquisition rates by GLH in young infected fields explain why tungro spreads very fast following the appearance of a few severely infected fields in asynchronous cropping areas. The role of infected ratoon fields as a source of infective migrants is limited because of the low population (see Figure 2) and low infectivity of GLH (Suwela et al., 1992).

This is in marked contrast with the situation in synchronous cropping areas where the vector sources for the subsequent crop consist mainly of ratoon and volunteer rice, provided that immigration of GLH from outside of the area is negligible. The means by which tungro occurs in small-scale synchronous cropping areas is not well understood. A likely cause of serious tungro occurrence is suggested from our experience in Bali. When transplanting in a few fields preceded the majority of fields in an area, tungro intensity was invariably far more serious in these early planted fields. If these fields were severely infected when young, tungro spread rapidly to the surrounding fields. Conversely, delayed transplanting of a small portion of an area did not lead to serious tungro infection in those fields if the delay was less than two weeks. This shows that serious infection is caused not by the variance in the transplanting date but by transplanting under conditions where the vector source/sink ratio is high. Analysis of the monthly area affected by tungro in Gianyar (this study) further supports this view.

For forecasting and monitoring tungro occurrence it may be instructive to see how parameters affect tungro incidence on the simple assumption that the two tungro viruses (rice tungro spherical virus and rice tungro bacilliform virus) are transmitted simultaneously. Suppose that tungro transmission by each generation of GLH occurs within a short period and that there is no overlap of generations.

Let D_i be GLH density, V_i be percentage of infective GLH and R_i be the population growth rate of GLH between generation i and $i+1$. The number of hills newly infected in generation i (H_i) is an increasing function of infective GLH density (D_iV_i) and, given that the overlap of infection is negligible, it may be assumed that H_i is proportional to D_iV_i.

Then $H_i = tD_iV_i$ where t is the transmission efficiency.

It is also assumed that, for $i \geq 1$, V_i increases with the increase in the number of hills infected before i at a constant rate of a. Suwela et al. (1992) give evidence for this assumption.

Numbers of newly infected hills by G_0, G_1 and G_2 are expressed as follows:

$$H_0 = tV_0D_0 \tag{1}$$

$$H_i = aH_0tD_i = at^2R_0V_0D_0^2 \tag{2}$$

$$H_2 = a(H_0 + H_1)tD_2 = at^2R_0R_1V_0D_0^2\,(1 + atR_0D_0) \tag{3}$$

There are important implications of the above equations. Firstly, consider the situation where H_2 is negligible, as is usual in asynchronous cropping areas. It is seen from Equations 1 and 2 that the amount of tungro transmission by G_1 relative to that of G_0, which is H_1/H_0, depends on R_0D_0, the population density of G_1, and does not depend on V_0; H_0 is proportional to both D_0 and V_0, while H_1 is a linear function of D_0^2 and V_0. Therefore, as G_1 density increases and H_1/H_0 becomes larger, the total infection (H_1+H_0) caused by G_0 and G_1 becomes more sensitive to D_0 than to V_0. In other words, fluctuations in the population density of immigrant GLH have greater influence on tungro occurrence than fluctuations in the infectivity of immigrant GLH. Note that this does not necessarily mean that fluctuation in the density of G_0 is the key factor causing fluctuations in tungro incidence because the coefficient variance is usually much larger in V_0 than in D_0.

Tungro transmission by G_2 may be significant in synchronous cropping areas where the highest population peak of GLH often comes in G_2. Provided that $H_1>>H_0$, it follows from Equation 3 that H_2 is proportional to D_0^3. Hence monitoring of migrant GLH density is particularly important in small-scale synchronous areas as far as virus sources exist.

SUMMARY

Field epidemiology of rice tungro disease and the population dynamics of its major vector, the green leafhopper (GLH), *Nephotettix virescens,* were studied in Bali, Indonesia, from 1987 to 1990 to develop forecasting technologies in asynchronous rice cropping areas where tungro is endemic.

- The highest population density of GLH in asynchronous areas occurred either in the immigrant or the first subsequent generation, unlike the populations in small-scale synchronous areas where the highest density was in the second generation.

- First-generation nymphs of GLH are the principal vectors for tungro transmission. Infection in seed-beds was negligible because of very low GLH density.

- Yield loss of an infected hill depended on the stage when the disease occurred and the location of the hill within the paddy field. First-generation nymphs of GLH which spread tungro to form patches of infected hills were most responsible for the yield loss.

- The infective vector index, expressed as the product of the number of GLH adults and the percentage of diseased hills with obvious symptoms two to five weeks after transplanting, was useful in predicting cumulative tungro infection at harvest.

- Regression analyses of six years of data on monthly tungro infection in the Gianyar Regency showed that the newly infected area at month t largely depended on the newly infected area at t-1, which was the major source of infective vectors, and the newly transplanted area at t-3, the major source of migrant vectors.

- The importance of monitoring migrant GLH density in forecasting tungro occurrence is discussed on both theoretical and empirical bases.

ACKNOWLEDGEMENTS

Thanks are due to J.M. Thresh, D.A. Andow and T.C.B. Chancellor for their valuable comments on earlier drafts of this paper and to the GLH/tungro study group of Indonesia–Japan Joint Programme on Food Crop Protection in Bali, Jakarta, Jatisari and Bogor for their help in the field work and discussions.

REFERENCES

ARYAWAN, G.N., GEDE, G.N. and SUZUKI, Y. (1993a) Population growth patterns of the green leafhopper, *Nephotettix virescens* (Distant) (Homoptera: Euscelidae), in small-scale synchronous and asynchronous rice fields. *Applied Entomology and Zoology*, **28**: 390–393.

ARYAWAN, G.N., WIDIARTA, N., SUZUKI, Y. and NAKASUJI, F. (1993b) Life table analysis of the green rice leafhopper, *Nephotettix virescens* (Distant) (Hemiptera: Cicadellidae), an efficient vector of rice tungro disease in asynchronous rice fields in Indonesia. *Researches on Population Ecology*, **35**: 31–43.

ASTIKA, N.S., SUWELA, N., ASTIKA, G.N. and SUZUKI, Y. (1992) Dependence of incubation period and symptoms of rice tungro disease (RTD) on infected stage in rice fields. *International Rice Research Newsletter*, **17**: 19–20.

CABUNAGAN, R.C. and HIBINO, H. (1989) Rice tungro (RTV) and its vector leafhopper development in synchronized-planting areas. *International Rice Research Newsletter*, **14**: 2.

CARIÑO, F.O., KENMORE, P.E. and DYCK, V.A. (1979) The FARMCOP suction sampler for hoppers and predators in flooded rice fields. *International Rice Research Newsletter*, **4**: 21–22.

ESTANO, D.B. and SHEPARD, B.M. (1989) Effect of roguing on rice tungro virus (RTV) incidence and rice yield. *International Rice Research Newsletter*, **14**: 22.

HINO, T., WATHANAKUL, L., NABHEERONG, N., SURIN, P., CHAIMONGKOL, U., DISTHAPORN, S., PUTTA, M., KERDCHOKCHAI, D. and SURIN, A. (1974) Studies on rice yellow leaf virus disease in Thailand. pp. 1–67. In: *Technical Bulletin* No. 7. Japan: Tropical Agriculture Research Center (TARC).

LOEVINSOHN, M.E. and ALVIOLA, A.A. (1991) Effect of asynchronized rice planting on vector abundance and tungro (RTD) infection. *International Rice Research Newsletter*, **16**: 20–21.

NARAYANASAMY, P. (1972) Influence of age of rice plant at the time of inoculation on the recovery of tungro virus by *Nephotettix impicticeps* (Ishihara). *Phytopathologische Zeitschrift*, **74**: 109–114.

SAMA, S., HASANUDDIN, A., MANWAN, I., CABUNAGAN, R.C. and HIBINO, H. (1991) Integrated rice tungro disease management in South Sulawesi, Indonesia. *Crop Protection*, **10**: 34–40.

SAWADA, H., KUSMAYADI, A., SUBUROTO, S.W.G., SUWARDIWIJAYA, E. and MUSTAGHFIRIN, (1993) Comparative analysis of population characteristics of the brown planthopper, *Nilaparvata lugens* Stål, between wet and dry rice cropping seasons in West Java, Indonesia. *Researches on Population Ecology*, **35**: 113–137.

SUWELA, N., ARYAWAN, G.N., ASTIKA, G.N. and SUZUKI, Y. (1992) Effect of rice stage and tungro (RTD) intensity on the infectivity of green leafhopper (GLH) in fields. *International Rice Research Newsletter*, **17**: 27.

SUZUKI, Y., ASTIKA, G.N., GEDE, G.N., WIDRAWAN, K.R., RAGA, N., YASIS and SOEROTO (1989) Field epidemiology of rice tungro disease and its management. pp. 32–34. In: *Proceedings of the Tenth National Congress of the Phytopathological Society of Indonesia.*

SUZUKI, Y., ASTIKA, G.N., WIDRAWAN, K.R., GEDE, G.N., RAGA, N. and SOEROTO (1992) Rice tungro disease vectored by the green leafhopper: its epidemiology and forecasting technology. *Japan Agriculture Research Quarterly*, **26**: 98–104.

SUZUKI, Y., RAINI, L.P.E. and ATMOIDJOJO, F.X.R. (1993) Forecasting rice tungro disease (RTD) occurrence in asynchronous rice planting areas on an empirical basis. *International Rice Research Notes*, **18**: 47.

TAIB, A.B. (1987) Status of rice pests and measures of control in the double cropping area of Muda irrigation scheme, Malaysia. pp. 107–115. In: *Tropical Agriculture Research Series* No. 20. Japan: Tropical Agriculture Research Center.

WIDIARTA, N., SUZUKI, Y., SAWADA, H. and NAKASUJI, F. (1990) Life tables and population dynamics of green leafhopper, *Nephotettix virescens* Distant (Hemiptera: Cicadellidae) in synchronized and staggered transplanting areas of paddy fields in Indonesia. *Researches on Population Ecology*, **32**: 319–328.

Paper 6

Role of vector control in management of rice tungro disease

T.C.B. CHANCELLOR,[1] K.L. HEONG[2] and A.G. COOK[1]

[1]*Natural Resources Institute, The University of Greenwich, Chatham Maritime, Chatham, Kent ME4 4TB, UK*
[2]*International Rice Research Institute, PO Box 933, 1099 Manila, Philippines*

INTRODUCTION

Rice tungro virus disease (RTVD) or rice tungro disease (RTD) has been described as a serious problem in rice in many countries in South and South-East Asia (Ling, 1972; Ou, 1985). The disease is caused by two viruses, rice tungro spherical virus (RTSV) and rice tungro bacilliform virus (RTBV) (Hibino *et al.*, 1978). Both viruses are transmitted by six leafhopper species, of which the most important are *Nephotettix virescens* (Distant) and *N. nigropictus* (Stål) (Hibino *et al.*, 1979). *N. virescens* is regarded as the main vector of tungro viruses as it is often more numerous in rice plantings than *N. nigropictus* and has a higher transmission efficiency (Hibino and Cabunagan, 1986). Another cicadellid, *Recilia dorsalis* (Motschulsky), is an inefficient vector of tungro viruses and population densities in rice crops are usually low, but it may be important in certain areas as relatively high numbers have been recorded on rice seed-beds and levees (Cook and Perfect, 1989).

The main strategy for control of RTD has been the use of vector-resistant varieties (Hibino *et al.*, 1990). Many such varieties were field-resistant to RTD at the time they were released but subsequently became infected as vector populations became adapted to them (Dahal *et al.*, 1990). Control of RTD through chemical control of the vectors has also been widely recommended, especially on susceptible varieties or on those where resistance 'breakdown' has occurred. This has led to concerns about the development of insecticide-resistance by leafhopper vectors and to the risks of pest resurgence following the widespread use of non-selective compounds (Litsinger, 1989). Furthermore, in addition to the financial outlay needed for chemical inputs, farmers face major health hazards associated with the long-term use of toxic compounds (Marquez *et al.*, 1992). The options available for vector control are discussed in relation to the ecology of the leafhopper vectors and the epidemiology of RTD. The need for a more targeted approach to vector control in RTD management is argued.

VIRUS/VECTOR INTERACTIONS AND RTD DYNAMICS

Both RTSV and RTBV are transmitted in a semi-persistent or transitory manner and vectors may remain viruliferous for up to four days (Chowdhury *et al.*, 1990). In laboratory studies, vectors have become infective after minimum feeding periods of 30 minutes on diseased rice plants (Rivera and Ou, 1965). Inoculation feeding times as short as 30 seconds have been recorded for successful virus transmission (Chowdhury *et al.*, 1990). Young rice plants are more easily infected with RTD than older plants and seedlings are highly susceptible (Ling, 1974). RTD vectors are found on seed-beds, sometimes in large numbers (Inoue *et al.*, 1975; Cook and Perfect, 1989). In Malaysia, high numbers of *N. virescens* were found on second-stage nurseries where seedlings are kept for 30–45 days before transplanting (Bottenberg *et al.*, 1990a). However, the precise role of seed-bed infection in tungro dynamics is not clear. A high incidence of infection in the seed-bed was reported in one trial where sowing was late and where seedlings were transplanted at 35 days old (Hino *et al.*, 1974). Mukhopadhyay *et al.* (1986) reported that seed-bed infection played an important role in RTD dynamics in outbreaks in West Bengal, India. Other workers found that seed-bed infection played a very limited role in RTD development in field trials (van Halteren, 1979; Tiongco *et al.*, 1993). From the evidence available, it seems likely that seed-bed infection is not important when seedlings are transplanted at 21 days of age or earlier, which is the common practice in most areas.

In transplanted rice, immigration of vector leafhoppers into fields begins shortly after transplanting and continues over a period of a few weeks (Inoue *et al.*, 1975). Tungro disease symptoms first appear in transplanted rice fields between 14 and 35 days after transplanting (DAT), although RTSV was detected by serology at 8 DAT in one field trial (Tiongco *et al.*, 1993). The latent period of infection in young rice plants was six to nine days in greenhouse studies (Rivera and Ou, 1965). Thus, it appears that tungro viruses are usually introduced into fields by viruliferous immigrant vectors after transplanting.

The importance of primary spread of RTD into plantings was demonstrated in recent field trials at the International Rice Research Institute (IRRI), where treatments with and without infected source plants were used. The results indicated that the inoculum was carried into the trial plots by viruliferous immigrants (Satapathy *et al.*, page 11). In a wet-season trial, when disease incidence was much higher than in the dry, there was also significant secondary spread of RTD within plantings of susceptible varieties as vectors acquired inoculum from infected plants and introduced it into previously healthy plants. The ability of tungro viruses to persist in the vector for several days and the short feeding periods needed for acquisition and transmission create the potential for rapid disease spread when conditions are favourable.

Nymphal stages of vector species can also transmit RTSV and RTBV (Ling, 1966). Anjaneyulu (1975) showed in field cage trials that nymphs spread viruses between plants, but their role in RTD spread has generally been regarded as minor due to their limited mobility. However, in an asynchronously planted area in Bali, Indonesia, RTD infections were highest when later stage *N. virescens* nymphs of the first field generation reached a peak (Suzuki *et al.*, 1992a).

CURRENT RTD VECTOR CONTROL OPTIONS

Susceptible plantings are at risk from RTD infection until the late vegetative stage, by which time plants become more difficult to infect and are less severely damaged when infected. Control measures aimed at reducing vector numbers should be targeted to be effective during the early critical period. The main options for RTD vector control are reviewed below.

Population regulation by natural enemies

Regulation of vector numbers by natural enemies may not prevent RTD from occurring in a rice crop, but it may help to reduce its incidence, particularly through limiting secondary disease spread. Little quantitative information is available, although high amounts of predation and parasitism of *Nephotettix* spp. have been recorded in rice crops.

Egg parasitism of *Nephotettix* spp. by dryinids (*Gonatocerus* spp.) and trichogrammatids (*Paracentrobia* spp.) varied from 45% to 100% in field populations in Sri Lanka (Fowler *et al.*, 1991). However, mortality from egg parasitism was inversely density dependent so that the efficiency of population regulation at higher pest levels would be reduced. Predation of *Nephotettix* eggs by a mirid (*Cyrtorhinus lividipennis* Reuter) was found to be high in Laguna Province, Philippines (Cook and Perfect, 1989). Cariño (1981) reported that nymphal predation by a veliid (*Microvelia douglasi atrolineata* Bergroth), and by spiders significantly reduced *Nephotettix* numbers in the field.

Surveys in Bali, Indonesia, showed that up to 29% of *N. virescens* adults were parasitized by pipunculids (Suzuki *et al.*, 1992b). In a study in Laguna Province, Philippines, parasitism by pipunculids reached a peak of 55% at one site (Pena and Shepard, 1986). Lycosids are the most effective group amongst a range of generalist spider predators that prey on leafhoppers and planthoppers, including *Nephotettix* species.

Together with appropriate cultural control measures, natural enemy regulation of RTD vectors is likely to have a beneficial effect in reducing tungro incidence. In practice, conservation of natural enemies is often difficult due to the widespread application by many rice farmers of broad-spectrum insecticides for controlling leaf-feeding insects (Heong *et al.*, 1992).

Cultural control

The intensification of rice production in Asia during the last few decades has involved the adoption of cultural practices such as sequential cropping which favour the increase of pest populations (Litsinger, 1989). The use of cultural measures to reduce insect pest numbers is not always compatible with agronomic practices recommended for the highest yields in intensive rice systems. Cultural control methods are often difficult to implement as, unlike insecticide applications, they do not provide farmers with dramatic evidence of vector control (Heinrichs, 1979). Furthermore, measures such as the manipulation of planting dates are often constrained by practical considerations such as the limited availability of water and labour (Loevinsohn, 1984). Nevertheless, various cultural control measures aimed at reducing vector numbers in rice crops are recommended. Plant spacing, planting date and a crop-free period are the main cultural practices which have been used.

In small-plot trials in India, RTD incidence was lower, and grain yields higher, in treatments with closer plant spacings compared with those with wider spacings, even though vector numbers per plant were not significantly different (Shukla and Anjaneyulu, 1981a). The reasons for this are not clear but it is possible that vectors move less when plants are closely spaced, resulting in less secondary spread of tungro viruses. This type of behaviour was seen with the corn leafhopper vector of corn stunt and maize rayado fino marafivirus (*Dalbulus maidis* Delong and Wolcott) in maize crops (Power, 1992).

The effect of different planting dates on RTD incidence was investigated by Shukla and Anjaneyulu (1981b) in small-plot trials. They recorded highest numbers of *N. virescens* and RTD incidence in later plantings and suggested that farmers in eastern India could avoid RTD problems by planting early in the *kharif* season and late in *rabi* when vector populations were low. In an integrated RTD management scheme developed in South Sulawesi, Indonesia, and implemented in 1983, planting dates were scheduled to precede the periods of peak vector populations (Sama *et al.*, 1991). Since the operation of the scheme RTD incidence has been low. However, it has been difficult to assess the effect of planting date in reducing RTD incidence as it is only one of four components adopted in the disease management scheme (Hasanuddin *et al.*, page 94).

Litsinger (1989) discussed the increase in abundance of *N. virescens* and two other major rice pests at IRRI following the shift from single to multiple cropping; he advocated a break between cropping seasons lasting more than the duration of one insect pest generation to prevent the continuous cycling of populations throughout the year which was made possible by overlapping crops. Loevinsohn (1984) had earlier argued that this could be achieved by the synchronous planting of early maturing varieties within clearly defined areas in an irrigation scheme; he suggested that the rice-free period would need to be implemented in over 90% of such areas which should be at least 314 ha in size for the plan to be effective.

A crop-free period has been implemented in the Muda irrigation scheme in Malaysia and synchronous planting is part of the integrated RTD management scheme in South Sulawesi. These approaches are also aimed at reducing the amount of RTD sources in an area and not exclusively at vector control. In an analysis of historical survey data from the Muda irrigation scheme, RTD incidence and vector abundance were shown not to be related to rice cropping pattern in two of the three seasons considered (Bottenberg *et al.*, 1990b). The number of RTD vectors was higher in more intensively cultivated and asynchronously planted rice areas in one season but the relationship was weak. Some researchers have argued that asynchrony helps to stabilize pest numbers in rice areas by maintaining populations of natural enemies throughout the year (Lim and Heong, 1977; Sawada *et al.*, 1993).

Chemical control

Insecticides have been more effective in controlling diseases caused by viruses that are transmitted in a persistent as opposed to a non-persistent manner (Walkey, 1985). Non-persistent and semi-persistent viruses can often be acquired and transmitted by insects during feeding periods shorter than the time required for insecticides to kill them. In some cases, insecticide applications have even contributed to virus disease spread as a result of changes in vector behaviour which they have induced (e.g. Shanks and Chapman, 1965). However, insecticides have been widely recommended for use against RTD vectors and many researchers have claimed success in using them to control the disease (Heinrichs, 1979). Almost all trials have been done in transplanted rice and some have included seed-bed treatments as well as treatments at or after transplanting. Results from these trials have not been consistent and an overall evaluation is difficult. However, some conclusions may be drawn from the available data.

Under conditions of high disease pressure, foliar insecticide applications which rely on contact action and which have low persistency are unlikely to be effective in reducing spread, even if they are applied several times each cropping season. This is because immigration of viruliferous vectors into a planting may continue over a long period. In addition, leafhopper eggs are oviposited in the tissues of the leaf sheath of rice tillers and hatch after about seven days (Cheng and Pathak, 1971). Unless insecticides with ovicidal properties are used, emerging nymphs may be unaffected by compounds with short residual activity. Foliar applications may be easily washed off rice plants after heavy rain and generally show poor selectivity, being highly toxic to natural enemies. Such applications may instead favour the build-up of leafhopper and planthopper populations due to reduced numbers of natural enemies.

Pathak *et al.* (1967) found that foliar sprays of carbaryl, DDT, endrin and malathion had limited residual activity against *Nephotettix* spp. in greenhouse tests and that weekly sprays of 0.04% endrin gave poor control of leafhoppers and planthoppers in the field. The synthetic pyrethroid cypermethrin was reported to give good control of RTD by some workers but this required frequent applications (Mochida *et al.*, 1986; Satapathy and Anjaneyulu, 1984). A further disadvantage of using cypermethrin is that sprays of this compound were reported to favour the population development of the brown planthopper, *Nilaparvata lugens* (Stål) (Heinrichs and Mochida, 1984; Schoenly *et al.*, 1996). Chemical application in trap crops was proposed to overcome some of these limitations (Saxena *et al.*, 1988). Results from small-plot trials showed that RTD incidence was significantly reduced when border rows of a susceptible variety were planted 15 days earlier than the main crop and then sprayed weekly with cypermethrin up to 60 DAT. It is questionable whether such an approach, which relies on diverting vectors onto the trap crop, would be feasible or effective on a larger scale under field conditions.

Granular applications of insecticides have given good vector control and reduced RTD incidence in a number of trials (van Halteren and Sama, 1974; Satapathy and Anjaneyulu, 1986). Granular formulations of carbofuran, isoprocarb and BPMC were more effective in reducing RTD incidence in the field than other compounds tested because of their rapid activity and long persistence (Satapathy and Anjaneyulu, 1986). Greenhouse studies conducted by the same workers showed that vector mortality reached 100% within 20 minutes of feeding on plants treated with the most effective compound, carbofuran. However, RTD incidence in a susceptible variety still reached 24% at 50 DAT when carbofuran was broadcast in the field (Table 1). Heinrichs *et al.* (1986) showed that the protection of susceptible varieties against RTD by broadcasting carbofuran granules in the field was not economically justified. Root-zone application of granular insecticides has been shown to be more effective than broadcasting as the uptake of the chemical is more efficient and it degrades more slowly (Satapathy and Anjaneyulu, 1989). The application technology for this method has not yet been developed. It is unclear whether the cost:benefit ratio of root zone application of granular insecticides at a field level is favourable.

In practice, farmers commonly apply foliar sprays which may be targeted against a range of pests and which are relatively cheap and easily available. In Leyte, Philippines, 89% of farmers interviewed used sprays and about 60% of them used two to four sprays in the wet season (Heong *et al.*, 1994). In the Mekong Delta in Vietnam, 96% of farmers sprayed their fields and the average number of sprays per farmer was seven per season (Heong *et al.*, 1994). At both sites, more than three quarters of the farmers applied their first spray in the first 30 days after crop establishment. In Leyte, although 25% of respondents cited RTD as their major pest problem, only 4.2% and 1.4% of the total number of chemical applications were targeted at control of leafhopper vectors and RTD, respectively. At the early and late tillering stages leaffolders, *Cnaphalocrocis medinalis* (Guenée), and other lepidopterous larvae were the

Table 1 Rice tungro disease (RTD) incidence in cv. Pankaj treated with insecticides in the field

Insecticide	Disease incidence Angular values* (% values) Days after transplanting					Rate of infection (*r*)
	22	29	36	43	50	
Carbofuran	2.5 (0.3)	13.7 (6)	21.9 (14)	24.0 (17)	29.2 (24)	0.079
Isoprocarb	3.3 (0.4)	15.1 (7)	26.8 (20)	33.0 (30)	39.4 (40)	0.106
BPMC	4.4 (0.6)	18.3 (10)	29.9 (25)	34.8 (33)	41.2 (43)	0.092
Diazinon	3.9 (0.5)	14.9 (7)	27.2 (21)	33.4 (30)	39.5 (41)	0.108
Carbaryl + gamma isomer of HCH	6.5 (1.3)	25.6 (19)	45.6 (51)	56.6 (70)	64.8 (82)	0.142
Phorate	5.5 (0.9)	23.9 (17)	40.4 (42)	55.4 (69)	64.2 (81)	0.146
Quinalphos	6.3 (1.2)	25.9 (19)	42.9 (46)	55.1 (67)	65.9 (83)	0.144
Control	9.6 (2.8)	28.3 (23)	54.3 (66)	76.6 (94)	90.0 (100)	0.287
LSD (*p* = 0.05)	2.7	3.3	4.7	4.2	3.7	
LSD (*p* = 0.01)	3.7	4.5	6.5	5.8	5.1	

* Mean of three replicates
BPMC = butyl phenyl methyl carbamate; HCH = hexachlorocyclohexane
Source: Satapathy and Anjaneyulu, 1986

45

main targets. At later growth stages sprays were directed against rice bugs, *Leptocorisa oratorius* (Fabricius). The most commonly used insecticides were the organochlorine endosulfan, which was also used for snail control, and the chlorinated hydrocarbons methyl parathion and monocrotophos.

As early as the 1950s, Broadbent (1957) argued that with virus diseases which are transmitted by mobile vectors, the effective use of insecticides is dependent on the attitude of farmers throughout the whole locality. Only through the co-operation of farmers over a large area could such diseases be controlled successfully. This is not a realistic objective under the conditions of small-scale farming in most irrigated and rainfed areas in South and South-East Asia. An illustration of the problem is provided by data collected on the re-invasion of previously sprayed rice areas by insect pests (Heong, 1991; Schoenly *et al.*, 1996). Immediately after an insecticide spray, substantial reductions in numbers of all guilds occurred but this was followed by recolonization and within a week densities were as high as in untreated plots. Species richness for all three guilds studied was also significantly reduced after each spray, but recovered within a few days of recolonization. The issue of vector mobility raises important questions about the design of insecticide trials (Broadbent, 1957). Plots need to be sufficiently large and have adequate separation distances between them to minimize inter-plot movement.

As indicated by several contributors to this volume, chemical control of vectors to reduce RTD incidence is recommended in most countries where the disease occurs. However, recommendations differ both between and within countries. Recommended chemical control measures from three research centres in India are summarized in Table 2. Protection of seedlings in nurseries is recommended by the first and third institutes, through a combination of granular and foliar applications, but at the third, recommendations are qualified according to the season and on vector numbers in the seed-bed. Insecticide applications in the field at or after transplanting are recommended by all three but the types of application, and the criteria for their use, differ. The concept of a threshold number of insects of two vectors per plant is used at Central Rice Research Institute (CRRI). These examples illustrate the widely varying approaches adopted for chemical control of RTD vectors. Little, however, is known about what farmers are doing or the effectiveness of the treatments applied.

Vector-resistant varieties

Seven genes for resistance to *N. virescens* have been identified, of which six are dominant and one recessive (Siwi and Khush, 1977; Karim and Pathak, 1982). Resistance is due to non-preference and antibiosis (Heinrichs, 1979). Feeding is predominantly in the xylem rather than in the phloem on resistant varieties (Auclair *et al.*, 1982). As the ability of *N. virescens* to transmit RTD viruses is related to the amount of feeding in the phloem, vector-resistant varieties are less easily infected and are a poor source of virus inoculum (Heinrichs and Rapusas, 1984). In greenhouse studies, varieties with different genes for resistance conferred different levels of resistance against *N. virescens* and this was inversely correlated with the amount of infection with RTD (Heinrichs and Rapusas, 1983).

Resistance breeding to RTD vectors has featured prominently at IRRI and in national research programmes. The initial objective was to provide an effective and relatively inexpensive way of controlling vector numbers; more recently there has been a realization of the need to avoid causing adverse effects on human health, natural enemies and the environment as a whole. A standardized method of screening germplasm to select for resistance to *N. virescens* was developed at IRRI and has been widely adopted elsewhere. Seedlings of test varieties and resistant and susceptible controls in seedboxes are infested with large numbers of *N. virescens*. The performance of the test varieties in comparison with the resistant and susceptible controls under heavy leafhopper infestation levels is assessed as a rapid means of selecting resistant germplasm. Promising lines or varieties are subjected to further tests to monitor vector feeding preference, survival and reproduction in order to evaluate the degree of resistance more accurately (Heinrichs and Rapusas, 1983).

Only IR22, of all the 'IR' varieties released for irrigated or rainfed lowland rice environments, had no vector resistance at the time of release. Four of the seven known genes for resistance to *N. virescens* have been incorporated into improved varieties (Khush, 1989). Such varieties showed field resistance to RTD at the time they were released but subsequently became infected after several seasons of intensive cultivation (Dahal *et al.*, 1990). This change in the field response to RTD of vector-resistant varieties was linked to a shift in the virulence of vectors to the varieties. Populations of *N. virescens* collected in Los Baños in 1986 fed mainly from the xylem and transmitted predominantly RTBV on IR54. Populations collected at the same location in 1987 and 1988 fed mainly from the phloem and efficiently transmitted RTSV and RTBV together to IR54 (Dahal *et al.*, 1990). The rotation of vector resistant and susceptible varieties practised in South Sulawesi, Indonesia, is an attempt to prevent or slow down the adaptation of vectors to resistant varieties (Sama *et al.*, 1991).

Table 2   Recommendations for chemical control of tungro vectors in India

Institute	Seed-bed	Transplanting
Directorate of Rice Research, Hyderabad	Carbofuran 3 g @ 30–35 kg/ha* or phorate 10 g @ 12–15 kg/ha* or seedlings; incorporated or broadcast 4–5 DAS	Carbofuran 3 g @ 25 kg/ha* or phorate 10 g @ 7.5 kg/ha* at 10 DAT
Source: Leaflet *Managing Tungro Disease in Rice*	Foliar spray at 15 and 25 DAS, depending upon GLH numbers	Further measures 'according to need'
Central Rice Research Institute, Cuttack	No insecticide application recommended	Threshold for insecticide application: 2 GLH per plant; Furadan 3 g @ 1.5 kg ai/ha or Ripcord 10% EC @ 0.01%/ha*
Source: Leaflet *Save Your Rice Crop from Tungro*		Repeat twice at 15-day intervals
Plant Virus Research Centre, Bidham Chandra Krishi Viswavidyalaya, Kalyani, West Bengal	Systemic insecticide as a prophylaxis e.g. Furadan 3 g @ 1.5 kg ai/ha or Foratox 10 g @ 1.75 kg ai/ha	Single application of insecticide e.g. an emulsifiable concentrate at *c*. 15 DAT
Source: Mukhopadhyay *et al.* (1986)	Second insecticide application at 4 days before transplanting: systemic granular compound or an emulsifiable concentrate such as quinalphos*†	

* Quantity of active ingredient in the formulation not stated.
† Recommendation dependent on season and on vector numbers in the seed-bed.
DAS = days after sowing; DAT = days after transplanting

There is evidence that certain 'IR' varieties were susceptible to populations of *N. virescens* in some locations at the time they were released (Karim and Pathak, 1982; Heinrichs *et al.*, 1986). This indicates that there are differences in virulence characteristics between field populations of *N. virescens*; these need to be considered in evaluating methodologies for screening germplasm for vector resistance.

Vector numbers and tungro incidence

One reason for the lack of consistency in recommendations for vector control may be that the quantitative relationship between vector numbers and RTD incidence is not clear. Outbreaks of RTD have often been associated with unusually high populations of vectors and occur more often in the wet season, when conditions are more favourable for vector population development, than in the dry season (Lim, 1972). However, large numbers of vectors do not necessarily lead to high RTD incidence, particularly in areas where disease sources are so few as to be limiting (Sogawa, 1976).

RTD incidence and vector numbers were correlated in only one of three seasons in studies done in the Muda irrigation scheme (Bottenberg *et al.*, 1990b); the authors suggested that when RTD sources are already present in an area, vector abundance may be less significant in determining disease incidence. An analysis of survey data in central Luzon, Philippines, where RTD outbreaks occur infrequently, revealed that epidemic years were mainly associated with the number of viruliferous vectors and that the total number of vectors was of secondary significance (Savary *et al.*, 1993). Data from two RTD-endemic areas included in the same study showed that the increase in disease incidence was associated with increasing vector numbers and with the number of viruliferous vectors. However, there were differences between the two sites in terms of the detailed interactions between the three variables.

It has been shown that RTD incidence may be high in a susceptible variety even when vector numbers are low. In experiments conducted using field cages, RTD incidence reached 72% with only one adult *N. virescens* per hill in the presence of an inoculum source (Shukla and Anjaneyulu, 1982). In a late-planted dry-season field trial at IRRI, RTD incidence in susceptible IR22 was 63% at 48 DAT

(Chancellor *et al.*, 1996). The peak adult vector population density was 2.1/hill at 36 DAT and nymphal populations were also low. Population development curves for nymphs and adults of the predominant vector species, *N. virescens*, are shown in relation to RTD incidence in Figure 1.

Outputs from a simulation model developed to help improve RTD risk assessment suggest that the balance between viruliferous and non-viruliferous immigrant vectors has a large impact on the rate of disease spread in a plot. A method of monitoring the inoculum pressure in a rice area was developed in Malaysia through the use of mobile nurseries to detect viruliferous vectors before planting, but there was no clear correlation with subsequent RTD incidence in the field (Chang, 1990). A similar type of approach was followed in field trials at IRRI where overflying vectors were collected in an upwardly directed light trap set at a height of 4.0 m and subsequently used in transmission tests with a susceptible variety (Chancellor *et al.*, 1997). The results showed that aerial populations of vectors were highest during the mid to late part of the wet season (June–October) and at the end of the dry season (January–April). Viruliferous vectors were collected during the same periods. The relationship between viruliferous vector numbers and onset of RTD in the field is currently being analysed.

In central Luzon, the number of vectors – and the proportion that were viruliferous – in the transition period from dry to wet seasons (May–July) was correlated with the incidence of RTD in the field from June–October (Ling *et al.*, 1983). For RTD forecasting, monitoring vector infectivity is not feasible in most rice-growing areas and information may be available too late to enable farmers to take appropriate control measures. In asynchronously planted areas in Bali, RTD incidence in the wet season was correlated with disease incidence in the second half of the dry season (Suzuki *et al.*, 1992a). Provided that RTD diagnosis is accurate and timely and that resources are adequate, field monitoring could be done by extension staff and used as a basis for recommendations to farmers.

CONCLUSIONS

In several countries in South and South-East Asia, RTD is endemic in certain areas and occurs sporadically, although sometimes on a larger scale, in non-endemic areas (Savary *et al.*, 1993). In the extensive areas of rice production where the disease occurs infrequently, or is unknown, the need to reduce vector populations on a routine basis may be questioned. Within endemic areas the incidence of RTD fluctuates from one year to the next. A need exists to identify the areas at risk from the disease and to target appropriate control measures accordingly. Currently, little information is available on the size of the area at risk of infection from RTD when a source of inoculum is identified in a locality. Therefore, it is difficult to assess the area over which protective measures should be applied. The role that vector control has in these cases should also be carefully appraised. There are practical difficulties in reducing vector numbers to sufficiently low densities to have a major impact on RTD incidence.

Vector-resistant varieties are likely to continue to be widely used during the next few years to protect crops against RTD. The incorporation of virus resistance or virus tolerance into such varieties is likely to prolong their useful life (Koganezawa and Cabunagan, page 54). Consideration needs to be given to the most effective way of deploying virus/vector-resistant varieties to maximize their impact. Due to limited availability or high cost, not all farmers have access to seed of resistant varieties. Some farmers may continue to grow susceptible varieties because of their superior eating quality or higher yield potential.

Farmers are likely to continue using insecticides to protect their rice crops, mainly because they perceive that they are needed. It is doubtful whether seed-bed protection as currently recommended, either routinely or when there is considered to be a threat of RTD, is necessary under most conditions but further research is needed to verify this. Insecticide applications are unnecessary and economically unjustifiable on vector-resistant varieties which remain field-resistant to RTD (Heinrichs *et al.*, 1986). On moderately resistant or susceptible varieties where there is a clearly identifiable risk of RTD, based on seasonal factors and disease incidence in the area, suitable insecticides may be a viable control option for farmers. Under such conditions, the requirements for effective control of RTD through insecticides are timely and appropriate application, fast knockdown and long persistence. However, these properties can also encourage secondary pest development (Heinrichs and Mochida, 1984). The safety of spray operators and of local communities as a whole in rural areas must be considered. Likely benefits from insecticide applications, in terms of a reduction in yield losses, must be balanced against the potential risks of secondary development of high populations of *Nilaparvata lugens*. Care

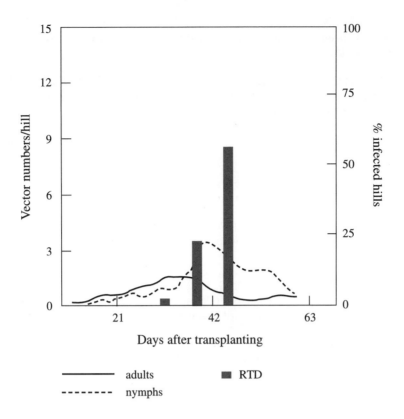

Figure 1 Populations of *Nephotettix virescens* adults and nymphs and rice tungro disease (RTD) incidence in the 1991 dry season on the IRRI farm, Los Baños, Philippines.

needs to be taken to avoid the emergence of vector resistance to chemical compounds. In Japan in the 1960s, *Nephotettix cincticeps* became resistant to organophosphate and carbamate compounds which had been heavily used (Heinrichs, 1979).

The use of cultural control measures such as adjustment of planting dates, synchronous planting and a crop-free period is possible only in certain areas. The limited availability of water and labour, and social and cultural factors have a major impact on rice cropping patterns. Further research is needed to determine whether asynchronous planting helps to stabilize, rather than to enhance, vector populations. Reviewing the available evidence in the literature, Way and Heong (1994) argued that asynchrony may in fact lead to higher stability, especially in rice ecosystems with high habitat diversity. One of the main arguments for synchrony is that RTD sources are substantially reduced at the time of planting, in the absence of standing rice crops. The issue of whether RTD inoculum sources can be reduced sufficiently under asynchronous conditions to minimize the threat of disease spread needs to be addressed.

ACKNOWLEDGEMENTS

We wish to thank all the staff in the Entomology and Plant Pathology Department at IRRI who have assisted in some of the studies described. The current research activities in RTD epidemiology and management referred to in this review were commissioned by the Natural Resources Institute as part of the Crop Protection Programme under the Overseas Development Administration's Renewable Natural Resources Research Strategy.

REFERENCES

ANJANEYULU, A. (1975) *Nephotettix virescens* (Distant) nymphs and their role in the spread of rice tungro virus. *Current Science (India)*, **44**: 357–358.

AUCLAIR, J.L., BALDOS, E. and HEINRICHS, E.A. (1982) Biochemical evidence for the feeding sites of the leafhopper, *Nephotettix virescens* within susceptible and resistant rice plants. *Insect Science and its Application,* **3**: 29–34.

BOTTENBERG, H., LITSINGER, J.A., BARRION, A.T. and KENMORE, P.E. (1990a) Presence of tungro vectors and their natural enemies in different rice habitats in Malaysia. *Agriculture, Ecosystems and Environment*, **31**: 1–15.

BOTTENBERG H., LITSINGER, J.A., LOEVINSOHN, M.E. and KENMORE, P.E. (1990b) Impact of cropping intensity and asynchrony on the epidemiology of rice tungro virus in Malaysia. *Journal of Plant Protection in the Tropics*, **7**: 1103–1116.

BROADBENT, L. (1957) Insecticidal control of the spread of plant viruses. *Annual Review of Entomology*, **2**: 339–354.

CARIÑO, F.O. (1981) *Role of natural enemies in population suppression and pest management of green rice leafhoppers*. PhD Thesis, University of the Philippines, Los Baños.

CHANCELLOR, T.C.B., COOK, A.G. and HEONG, K.L. (1996) The within-field dynamics of rice tungro disease in relation to the abundance of its major leafhopper vectors. *Crop Protection*, **15**: 439–449.

CHANCELLOR, T.C.B., COOK, A.G., HEONG, K.L. and VILLAREAL, S. (1997) The flight activity and infectivity of the major leafhopper vectors (Hemiptera: Cicadellidae) of rice tungro viruses in an irrigated rice area in the Philippines. *Bulletin of Entomological Research*, **87**: 247–258.

CHANG, P.M. (1990) The leafhopper vectors of tungro in Malaysia. Paper presented at a *Workshop on Penyakit Merah Virus in Telok Chengai, Kedah, 14–16 May 1990*.

CHENG, C.H. and PATHAK, M.D. (1971) Bionomics of the rice green leafhopper, *Nephotettix impicticeps* Ishishara. *Philippine Entomologist*, **2**: 67–74.

CHOWDHURY, A.K., TENG, P.S. and HIBINO, H. (1990) Retention of tungro-associated viruses by leafhoppers and its relation to rice cultivars. *International Rice Research Newsletter*, **15**: 31.

COOK, A.G. and PERFECT, T.J. (1989) Population dynamics of three leafhopper vectors of rice tungro viruses, *Nephotettix virescens* (Distant), *N. nigropictus* (Stål) and *Recilia dorsalis* (Motschulsky) (Hemiptera: Cicadellidae), in farmers' fields in the Philippines. *Bulletin of Entomological Research*, **79**: 437–451.

DAHAL, G., HIBINO, H., CABUNAGAN, R.C., TIONGCO, E.R., FLORES, Z.M. and AGUIERO, V.M. (1990) Changes in cultivar reactions to tungro due to changes in 'virulence' of the leafhopper vector. *Phytopathology*, **80**: 659–665.

FOWLER, S.V., CLARIDGE, M.F., MORGAN, J.C., PERIES, I.D.R. and NUGALIYADDE, L. (1991) Egg mortality of the brown planthopper, *Nilaparvata lugens* (Homoptera: Delphacidae) and green leafhoppers, *Nephotettix* spp. (Homoptera: Cicadellidae), on rice in Sri Lanka. *Bulletin of Entomological Research*, **81**: 161–167.

VAN HALTEREN, P. (1979) The insect pest complex and related problems of lowland rice cultivation in South Sulawesi, Indonesia. *Mededelingen Landbouwhogeschool Wageningen*, **79**.

VAN HALTEREN, P. and SAMA, S. (1974) The tungro virus disease in Sulawesi. pp. 175–184. In: *Agricultural Co-operation between Indonesia and the Netherlands, Research Reports* (1968–1974).

HEINRICHS, E.A. (1979) Control of leafhopper and planthopper vectors of rice viruses. pp. 529–560. In: *Leafhopper Vectors and Plant Disease Agents*. MARAMAROSCH, K. and HARRIS, K.F. (eds). New York: Academic Press.

HEINRICHS, E.A. and MOCHIDA, O. (1984) From secondary to major pest status: the case of insecticide-induced rice brown planthopper, *Nilaparvata lugens*, resurgence. *Protection Ecology*, **7**: 201–218.

HEINRICHS, E.A. and RAPUSAS, H. (1983) Correlation of resistance to the green leafhopper, *Nephotettix virescens* (Homoptera: Cicadellidae) with tungro virus infection in rice varieties having different genes for resistance. *Environmental Entomology*, **12**: 201–205.

HEINRICHS, E.A. and RAPUSAS, H. (1984) Feeding, development, and tungro virus transmission by the green leafhopper, *Nephotettix virescens* (Distant) (Homoptera: Cicadellidae) after selection on resistant rice cultivars. *Environmental Entomology,* **13**: 1074–1078.

HEINRICHS, E.A., RAPUSAS, H.R., AQUINO, G.B. and PALIS, F. (1986) Integration of host plant resistance and insecticides in the control of *Nephotettix virescens* (Homoptera: Cicadellidae) a vector of rice tungro virus. *Journal of Economic Entomology,* **79**: 437–443.

HEONG, K.L. (1991) Management of the brown planthopper in the tropics. In: *Proceedings of a Symposium on Migration and Dispersal of Agricultural Insects, Tsukuba, Japan, 25–28 September 1991.*

HEONG, K.L., ESCALADA, M.M. and LAZARO, A.A. (1992) *Pest Management Practices of Rice Farmers in Leyte, Philippines. Survey Report.* Los Baños, Philippines: International Rice Research Institute.

HEONG, K.L., ESCALADA, M.M. and VO MAI (1994) An analysis of insecticide use in rice: case studies in the Philippines and Vietnam. *International Journal of Pest Management,* **40**: 173–178.

HIBINO, H. and CABUNAGAN, R.C. (1986) Rice tungro-associated viruses and their relations to host plants and vector leafhoppers. pp. 173–182. In: *International Symposium on Virus Diseases of Rice and Leguminous Diseases in the Tropics.* Tropical Agricultural Research Series 19.

HIBINO, H., DAQUIOAG, R.D., MESINA, E.M. and AGUIERO, V.M. (1990) Resistances in rice to tungro-associated viruses. *Plant Disease,* **74**: 923–926.

HIBINO, H., ROECHAN, M. and SUDARISMAN, S. (1978) Association of two types of virus particles with penyakit habang (tungro disease) of rice in Indonesia. *Phytopathology,* **68**: 1412–1416.

HIBINO, H., SALEH, N. and ROECHAN, M. (1979) Transmission of two kinds of rice tungro-associated viruses by insect vectors. *Phytopathology,* **69**: 1266–1268.

HINO, T., WATHANAKUL, L., NABHEERONG, N., SURIN, P., CHAIMONGKOL, U., DISTHAPORN, S., PUTTA, M., KERDCHOKCHAI, D. and SURIN, A. (1974) Studies on rice yellow orange leaf virus disease in Thailand. pp 1–67. In: *Technical Bulletin* No. 7. Japan: Tropical Agricultural Research Center (TARC).

INOUE, H., RUAYAREE, S., HENGSAWAD, C., HENGSAWAD, V. and PATIRUPANUSARA, P. (1975) *Studies on Rice Green Leafhopper in Thailand in Relation to Yellow Orange Leaf Virus Disease.* Final Report of the Joint Research Work between Thailand and Japan. [mimeo.]

KARIM, A.N.M.R. and PATHAK, M.D. (1982) New genes for resistance to green leafhopper, *Nephotettix virescens* (Distant) in rice *Oryza sativa* L. *Crop Protection,* **1**: 483–490.

KHUSH, G.S. (1989) Multiple disease and insect resistance for increased yield stability in rice. In: *Progress in Irrigated Rice Research.* Los Baños, Philippines: International Rice Research Institute.

LIM, G.S. (1972) Studies on penyakit merah disease of rice III. Factors contributing to an epidemic in North Krian, Malaysia. *Malaysian Agricultural Journal,* **48**: 278–294.

LIM, G.S. and HEONG, K.L. (1977) *Habitat Modification for Regulating Pest Populations of Rice in Malaysia.* Malaysia Agricultural Research and Development Institute Report, No. 50. Malaysia Agricultural Research and Development Institute.

LING, K.C. (1966) Nonpersistence of the tungro virus of rice in its leafhopper vector, *Nephotettix impicticeps. Phytopathology,* **56**, 1252–1256.

LING, K.C. (1972) *Rice Virus Diseases.* Los Baños, Philippines: International Rice Research Institute.

LING, K.C. (1974) The capacity of *Nephotettix virescens* to infect rice seedlings with tungro. *Philippine Phytopathology,* **10**: 42–49.

LING, K.C., TIONGCO, E.R and FLORES, Z.M. (1983) Epidemiological studies of rice tungro. pp. 249–257. In: *Plant Virus Epidemiology*. PLUMB, R.T. and THRESH, J.M. (eds). Oxford: Blackwell Scientific Publications.

LITSINGER, J.A. (1989) Second generation insect pest problems on high yielding rices. *Tropical Pest Management*, **35**: 235–242.

LOEVINSOHN, M.E. (1984) *The Ecology and Control of Rice Pests in Relation to the Intensity and Synchrony of Cultivation*. PhD Thesis, University of London.

MARQUEZ, C.B., PINGALI, P.L. and PALIS, F.G. (1992) Farmer health impact of long term pesticide exposure — a medical and economic analysis in the Philippines. Paper presented at a *Workshop on Measuring the Health and Environmental Effects of Pesticides, Bellagio, Italy, 30 March–3 April 1992*.

MOCHIDA, O., VALENCIA, S.L. and BASILIO, R.P. (1986) Chemical control of green leafhoppers to prevent virus diseases, especially on susceptible/intermediate rice cultivars in the tropics. pp. 195–208. In: *International Symposium on Virus Diseases of Rice and Leguminous Diseases in the Tropics*. Tropical Agricultural Research Series, no. 19.

MUKHOPADHYAY, S., CHOWDHURY, A.K. and CHAKRABARTY, S.K. (1986) Epidemiology and control of rice tungro virus disease in West Bengal. pp. 142–153. In: *Proceedings of the National Seminar on Ricehoppers, Hopperborne Viruses and their Integrated Management, West Bengal*.

OU, S.H. (1985) *Rice Diseases*. Wallingford, UK: Commonwealth Agricultural Bureau.

PATHAK, M.D., VEA, E. and JOHN, V.T (1967) Control of insect vectors to prevent virus infection of rice plants. *Journal of Economic Entomology*, **60**: 218–225.

PENA, N. and SHEPARD, M. (1986) Seasonal incidence of parasitism of brown planthoppers, *Nilaparvata lugens* (Homoptera: Delphacidae), green leafhoppers, *Nephotettix* spp. and whitebacked planthoppers, *Sogatella furcifera* (Homoptera: Cicadellidae) in Laguna Province, Philippines. *Environmental Entomology*, **15**: 263–267.

POWER, A.G. (1992) Host plant dispersion, leafhopper movement and disease transmission. *Ecological Entomology*, **17**: 63–68.

RIVERA, C.T. and OU, S.H. (1965) Leafhopper transmission of 'tungro' disease of rice. *Plant Disease Reporter*, **49**: 127–131.

SAMA, S., HASANUDDIN, A., MANWAN, I., CABUNAGAN, R.C. and HIBINO, H. (1991) Integrated management of rice tungro disease in South Sulawesi, Indonesia. *Crop Protection*, **10**: 34–40.

SATAPATHY, M.K. and ANJANEYULU, A. (1984) Use of cypermethrin, a synthetic pyrethroid, in the control of rice tungro virus disease and its vector. *Tropical Pest Management*, **30**: 170–178.

SATAPATHY, M.K. and ANJANEYULU, A. (1986) Prevention of rice tungro virus disease and control of the vector with granular insecticides. *Annals of Applied Biology*, **108**: 503–510.

SATAPATHY, M.K. and ANJANEYULU, A. (1989) Effect of root zone placement of granular insecticides for tungro prevention and its control. *Tropical Pest Management*, **35**: 51–56.

SAVARY, S., FABELLAR, N., TIONGCO, E.R. and TENG, P.S. (1993) A characterization of rice tungro epidemics in the Philippines from historical survey data. *Plant Disease*, **77**: 376–382.

SAWADA, H., KUSMAYADI, A., SUBROTO, S.W.G., SUWARDIWIJA,YA E. and MUSTAGHFIRIN (1993) Comparative analysis of population characteristics of the brown planthopper, *Nilaparvata lugens* Stål, between wet and dry rice cropping seasons in West Java, Indonesia. *Researches in Population Ecology*, **35**: 113–137.

SAXENA, R.C., JUSTO, H.D. JR. and PALANGINAN, E.L. (1988) Trap crop for *Nephotettix virescens* (Homoptera: Cicadellidae) and tungro management in rice. *Journal of Economic Entomology,* **81**: 1485–1488.

SCHOENLY, K.G., COHEN, J.E., HEONG, K.L., ARIDA, G.S., BARRION, A.T. and LITSINGER, J.A., (1996) Quantifying the impact of insecticides on food web structure of rice-arthropod populations in a Philippine farmers' irrigated field: a case study. pp. 343–351. In: *Food Webs: Integration of Patterns and Dynamics.* POLIS, G.A. and WINEMILLER, K.O., (eds). New York: Chapman and Hall.

SHANKS, C.H. JR. and CHAPMAN, R.K. (1965) The effects of insecticides on the behaviour of the green peach aphid and its transmission of potato virus Y. *Journal of Economic Entomology,* **58**: 79–83.

SHUKLA, V.D. and ANJANEYULU, A. (1981a) Plant spacing to reduce rice tungro incidence. *Plant Disease,* **65**: 584–586.

SHUKLA, V.D. and ANJANEYULU, A. (1981b) Adjustment of planting date to reduce rice tungro disease. *Plant Disease,* **65**: 409–411.

SHUKLA, V.D. and ANJANEYULU, A. (1982) Effects of numbers of leafhoppers and amount and source of virus inoculum on the spread of rice tungro. *Journal of Plant Diseases and Protection,* **89**: 325–331.

SIWI, B.H. and KHUSH, G.S. (1977) New genes for resistance to the green leafhopper in rice. *Crop Science,* **17**: 17–20.

SOGAWA, K. (1976) Rice tungro virus and its vectors in tropical Asia. *Review of Plant Protection Research,* **9**: 21–46.

SUZUKI, Y., ASTIKA, G.N., WIDRAWAN, K.R., GEDE, G.N., RAGA, N. and SOEROTO (1992a) Rice tungro disease transmitted by the green leafhopper: its epidemiology and forecasting technology. *Japanese Agricultural Research Quarterly,* **26**: 98–104

SUZUKI, Y., RAGA, N., NURUSANTI, S. and WIDRAWAN, K.R. (1992b) Effects of pipunculid parasitism on the population dynamics of the green rice leafhopper, *Nephotettix virescens* Distant. *Proceedings of the Association for Plant Protection of Kyushu,* **38**: 73–77.

TIONGCO, E.R., CABUNAGAN, R.C., FLORES Z.M., HIBINO, H. and KOGANEZAWA, H. (1993) Serological monitoring of rice tungro disease development in the field: its implication in disease management. *Plant Disease,* **77**: 877–882.

WALKEY, D.G.A. (1985) *Applied Plant Virology,* London: Heinemann.

WAY, M.J. and HEONG, K.L. (1994) The role of biodiversity in the dynamics and management of insect pests of tropical irrigated rice – a review. *Bulletin of Entomological Research,* **84**: 567–587.

Paper 7

Resistance to rice tungro virus disease

H. KOGANEZAWA* and R.C. CABUNAGAN

International Rice Research Institute, PO Box 933, 1099 Manila, Philippines

INTRODUCTION

Since rice tungro virus disease (RTVD), commonly known as tungro or rice tungro disease, first attracted attention as a major disease of rice in the Philippines in the 1960s, various control strategies have been proposed and implemented. Among them, the use of resistant varieties is the most convenient, effective and least expensive. For many years, efforts have been made to screen rice germplasm for tungro resistance at the International Rice Research Institute (IRRI) and by several national agricultural research systems. Many rice varieties showing field resistance to tungro that is now known to be due to resistance to the main leafhopper vector (*Nephotettix virescens*) were bred and released. However, breakdown of resistance occurred after several consecutive seasons of intensive cultivation of initially resistant cultivars (Inoue and Ruay-Aree, 1977; Dahal *et al.*, 1990). Subsequent research has shifted to virus resistance which may be more effective and durable. This paper considers the types of virus resistance, sources of resistance and their application in breeding virus-resistant varieties.

SCREENING METHODS

Screening can be done either in the field or greenhouse. The mass screening method for greenhouse-testing of resistance to tungro at IRRI was developed in 1964, modified in 1974 (Ling, 1974) and revised in 1987. Since 1991, the 'seedbox in water tray' method of Saxena *et al.* (1991) has been used for primary mass screening in the greenhouse. In this method, 20 seeds per variety are sown in a line in a seedbox. Seven days after sowing, the seedbox is placed in a water tray and covered by a screen cage into which viruliferous vectors are released. Two and half hours later, the tray is filled with water, driving the insects onto the screen. When the seedlings are totally submerged, the seedbox is gently removed. One month after inoculation, the disease severity index of individual plants is scored using the system of Hasanuddin *et al.* (1988):

1 = no symptoms
3 = 10% reduction in plant height
5 = 11–30% reduction with no distinct leaf discolouration
7 = 31–50% reduction with yellow or yellow-orange leaf discolouration
9 = >50% reduction with yellow or yellow-orange discolouration

The average score indicates the degree of resistance. However, the reaction of some varieties does not fit the above categories. For example, cv. ARC7007 does not express distinct leaf discolouration but shows more than 30% height reduction.

The results from mass screening are incorporated into a database at IRRI. The selected varieties which have low average scores are tested by test-tube inoculation using three viruliferous insects per plant and a 24-hour inoculation access period. One month after inoculation, plants are tested separately by enzyme-linked immunosorbent assay (ELISA) for each of the viruses associated with tungro disease: rice tungro spherical virus (RTSV) and rice tungro bacilliform virus (RTBV). About 16 000 accessions, which is *c.* 20% of the total in the International Rice Germplasm Centre (IRGC) at IRRI, were tested before 1989. Analysis revealed that most of the resistant varieties originated in south Asia, particularly in Bangladesh, India and Pakistan (Cabunagan and Koganezawa, 1993). The above mass screening method was later used to evaluate untested varieties originating from Bangladesh, Pakistan and Sri Lanka; of 8000 or so accessions, *c.* 10% showed resistance to tungro.

---

*Present address: Shikoku National Agricultural Experiment Station, Ministry of Agriculture, Forestry and Fisheries, 1-3-1, Senyu, Zentsuji, Kagawa 765, Japan.

RESISTANCE TO THE LEAFHOPPER VECTOR

Resistance to the leafhopper vector occurs commonly in rice germplasm. Four of the seven known genes for vector resistance have been incorporated into improved varieties (Khush, 1989). Most IRRI crosses made after 1969 had at least one parent with resistance derived from Ptb18, Gam Pai 30-12-15 or Ptb33: varieties which originated in India, Thailand and India, respectively. All IR varieties other than IR22 were rated as having at least some resistance to green leafhopper at the time of release. In India, rice cultivar Vikramarya is recommended as a tungro-resistant variety — it seems to have only vector resistance. Such cultivars escape tungro infection in the field under light to moderate tungro and vector pressure (Heinrichs and Rapusas, 1983; Hibino et al., 1987). Under field conditions, vectors may be unable to settle on vector-resistant varieties for sufficiently long to acquire virus particles from and inoculate them to vascular tissue.

So far, introduction of vector-resistant varieties has been a major component of integrated tungro control schemes in various regions. In the Muda area of Malaysia, vector-resistant or moderately resistant varieties such as IR42, MR71, MR73, MR77 and MR88 have been widely grown. In South Sulawesi, Indonesia, rice cultivars were classified into four groups, represented by IR26, IR42, IR54 and Pelita, and based on the type of vector resistance present. Varieties of the different categories are deployed in an appropriate rotation cycle depending on the tungro situation and planting season in each area (Manwan et al., 1985; Sama et al., 1991). In these areas, tungro incidence declined drastically after the introduction of vector-resistant varieties. Although other components of the control strategy might have contributed to the tungro reduction, there is little doubt that the introduction of vector-resistant cultivars reduced tungro incidence.

RESISTANCE TO RICE TUNGRO SPHERICAL VIRUS

Many varieties have been reported as resistant to RTSV infection (Hibino et al., 1990). In studies at IRRI, 508 IRGC accessions which were selected by primary mass screening were evaluated: 115 accessions showed RTSV infection rates of below 10%; 76 accessions were not infected with RTSV in the two trials (Cabunagan et al., 1993b). Of the 100 accessions with RTSV resistance listed in Table 1 36 accessions also have vector resistance. The RTSV infection rate of vector-resistant varieties is always low using the IRRI inoculation method and such varieties are not necessarily resistant to RTSV.

Also at IRRI, an Indian variety, TKM6, was used as a source of resistance to stem borers, bacterial leaf blight and tungro. TKM6 was crossed with a selection from Peta × Taichung Native 1 (TN1); IR20 was selected from the progeny and released in 1969 and was later found to be resistant to RTSV infection. IR26, IR30 and IR40, which also have TKM6 in their parentage, all show resistance to RTSV (Hibino et al., 1988). IR20 was grown in 1973 and 1974 by farmers, but was soon replaced because of its susceptibility to brown planthopper (Nilaparvata lugens) and poor grain quality. The Malaysian variety MR81 was developed from a cross between MR24 and IR36, both of which are susceptible to tungro; MR81 is resistant to RTSV. Recessive resistance genes from Pankhari 203 (through Sri Malaysia II which is a parent of MR24) and IR36 may have been combined together in MR81 (Imbe and Habibuddin, 1989) which is currently recommended for tungro-prone areas of Malaysia.

RTSV-resistant varieties show high infection rates with tungro disease in artificial inoculation tests because of their susceptibility to RTBV infection, but disease development is expected to be slow in fields (Sama et al., 1991). This is because RTBV cannot be acquired from plants infected with this virus alone, unless the vectors have had prior access to a source of RTSV (Hibino, 1983). When the RTSV-resistant variety IR26 was exposed to infection in the field, the distribution of infected plants was scattered (Cabunagan et al., 1989; Satapathy et al., page 11). By contrast, infected plants were more numerous and appeared in patches in the comparable plots of susceptible varieties in which secondary spread occurred.

Tungro virus strains were first reported in India (Anjaneyulu and John, 1972) and the Philippines (Rivera and Ou, 1967) – before the discovery that tungro is caused by two distinct viruses. Dahal et al. (1992) compared varietal reactions to three isolates from the Philippines, Malaysia and India, and demonstrated variation in symptom severity and transmission profile among the three isolates.

Table 1 Rice varieties showing resistance to infection with rice tungro spherical virus (RTSV)

Accession number	Variety	Accession number	Variety
177	Adday Sel.*	26789	Shalya†
180	Adday Local Sel.*	26791	Sham Rosh*
4021	Binicol†	26813	Gogoi*
5999	Pankhari 203†	27529	Bhoilush†
7366	PL 184675-2†	27779	Bara Pashawari 390
8261	Padi Kasalle	27781	Bara 143†
11062	G 378	27787	Basmati Nahan 381
11751	Habiganj DW 8	27798	Basmati 1
12203	ARC 6064†	27799	Basmati 43 A†
12274	ARC 6561	27800	Basmati 93†
12310	ARC 7007†	27803	Basmati 107†
12428	ARC 10312	27804	Basmati 113†
12437	ARC 10343	27805	Basmati 122†
14504	IR580 420-1-1-2	27814	Basmati 208†
14527	Barah	27818	Basmati 242
14649	Gendjah Melati	27821	Basmati 370 A†
14703	CPA 86805-2†	27828	Basmati 376†
15769	Lawangeen†	27829	Basmati 377
16680	Utri Merah*	27830	Basmati 388
16684	Utri Rajapan	27832	Basmati 405
19680	ARC 10963	27833	Basmati 406
20600	ARC 7321	27835	Basmati 427†
21164	ARC 10980	27836	Basmati 433
21310	ARC 11315	27856	Begumi 302
21337	ARC 11346	27869	Chahora 144
21342	ARC 11353	27870	Chahora 148
21473	ARC 11554†	27872	Chahora 292
21474	ARC 11555	27873	Chahora 382
21745	ARC 11920*	27916	Dhanlu 254
21958	ARC 12170	27943	Hansraj 54†
22176	ARC 12596	27946	Hansraj 62
22199	ARC 12620	27947	Hansraj 189
22215	ARC 12636	27948	Hansraj 197
22307	ARC 12746	27951	Hansraj 365 A
22331	ARC 12778	28102	P 590
26253	Nep Bap*	28320	Toga 286 A†
26295	Bale Betor†	28341	9†
26316	Birpala†	28450	361†
26410	Pala Bhir*	28522	Gundrikbhog
26418	Shada Muta†	28867	AUS 4
26495	Konek Chul*	31746	Bish Katari†
26527	Shuli 2	36731	Firro E(1)
26560	Baharat†	37215	Matichakma
26582	Buchi 2	37337	Urman Sardar
26622	Gia Dhan†	37430	Ghigos
26633	Gurdoi†	37482	Kanakchul
26663	Kaisha Binni†	37488	Kashiabinni†
26703	Kurki†	37491	Katijan†
26715	Lao Bhug†	37761	Maliabhangor 1096*
26784	Sakor†	49996	Ovarkondoh*

* Cultivars showing resistance to both RTSV strains A and Vt6.

† Apparent resistance to RTSV may be due to vector resistance.

The variety TKM6 reacted differently to RTSV isolates from India and the Philippines. In the Philippines, a virulent strain of RTSV designated Vt6 was found infecting rice at Midsayap, North Cotabato, on the southern island of Mindanao. This strain could readily infect TKM6 which is highly resistant to the IRRI strain of RTSV designated 'A'. Rice accessions previously identified as resistant to strain A were tested against Vt6. All the varieties tested were highly resistant to strain A, whereas their reaction to strain Vt6 varied from highly resistant (0–10% infection) to very susceptible. Twenty-nine varieties had an infection rate >60%, comparable to the susceptible check, TN1. These results indicate that RTSV resistance may break down when used widely, as already reported with vector resistance. Eleven accessions, however, were highly resistant (0–10% infection) to both strains (Table 1) and may be used as sources of resistance genes to both strains of RTSV (Cabauatan et al., 1995).

RESISTANCE TO RICE TUNGRO BACILLIFORM VIRUS

Currently there is no improved rice variety that is resistant to RTBV infection, although Hibino et al. (1990) demonstrated that some rice cultivars were resistant to infection with both RTBV and RTSV. To check their reports, we re-tested these accessions, but none showed resistance to infection with RTBV. Therefore, the 508 entries which showed a resistance reaction in mass screening were re-tested: only 10 accessions showed infection rates of <30%. Re-testing these 10 accessions showed that only ARC 11554 and Katijan had a low infection rate of RTBV. Both varieties are resistant to green leafhopper and in other tests sometimes showed a high infection rate with RTBV. ARC 11554 is susceptible to green leafhopper in India and showed an intermediate reaction (30–60% infection) to RTBV (International Rice Research Institute, 1993). These facts indicate that the resistance of ARC11554 to RTBV infection is mainly due to vector resistance. However, other vector-resistant varieties were infected with RTBV at a high rate in the same experiments, indicating that the vector resistance of ARC11554 is somewhat different from that of other varieties.

Table 2 Rice varieties showing tolerance to infection with rice tungro bacilliform virus (RTBV)

Accession number	Variety	Accession number	Variety
177	Adday Sel.	21956	ARC 12168
180	Adday Local Sel.	21958	ARC 12170
3707	Andi fr N. Pokhara	22176	ARC 12596
4021	Binicol	22199	ARC 12620
5346	Seratus Hari T 36	22215	ARC 12636
7366	P1 184675-2	22309	ARC 12748
11751	Habiganj DW 8	22331	ARC 12778
12203	ARC 6064	26418	Shada Muta
12207	ARC 6080	26468	Hansa
12274	ARC 6561	26494	Kola Mona
12310	ARC 7007	26495	Konek Chul
12428	ARC 10321	26527	Shuli 2
12437	ARC 10343	26582	Buchi 2
14527	Barah	26622	Gia Dhan
15769	Lawangeen	26663	Kaisha Binni
16602	Tjempo Kijik	26682	Kaisha Binni
16680	Utri Merah	26703	Kurki
16684	Utri Rajapan	26772	Pura Binni
17204	Balimau Putih	26776	Raja Mun
17292	Betrik	26791	Sham rosh
19675	ARC 5905	226831	Gogoj
19680	ARC 10963	27821	Basmati 370A
20533	ARC 7140	27827	Basmati 375A
20600	ARC 7321	27828	Basmati 376
21337	ARC 11346	31746	Bish Katari
21342	ARC 11353	37430	Ghigos
21344	ARC 11355	37473	Kalakura
21473	ARC 11554	37488	Kashia Binni
21474	ARC 11555	37491	Katijan
21476	ARC 11558	37509	Kushiyari
21569	ARC 11698	49996	Ovarkondoh
21745	ARC 11920		

Although sources for resistance to RTBV infection have not been found in rice germplasm, several varieties (Table 2) show symptomatic resistance or tolerance (Hasanuddin and Hibino, 1989; Hibino *et al.*, 1990). The term 'tolerance' is applied here to the varieties that show no or mild symptoms and no marked yield loss when infected, regardless of virus concentration.

Multiplication of RTBV is suppressed in some tolerant varieties such as Utri Merah, Balimau Putih and Utri Rajapan, and virus concentration is too low to be detected by ELISA (Takahashi *et al.*, 1993; Cabunagan *et al.*, 1993a). These varieties seem to be good sources for resistance because virus spread in fields is expected to be slow. Among varieties so far tested in several regions, Utri Merah, which originated in Indonesia, always showed tolerance. Efforts to transfer the tolerance gene from Utri Merah to improved varieties are being made at IRRI.

CONCLUSION

Breeding of resistant varieties will continue to be a main objective in tungro management research. One of the current goals of the breeding programme at IRRI is to combine vector resistance with RTBV tolerance from Utri Merah. However, the tolerance is reported to be governed by polygenes (Shahjahan *et al.*, 1990) and it will take some time to obtain elite lines tolerant to RTBV. As for incorporation of RTSV resistance to improved varieties, it is necessary to select as donor varieties those which will be effective throughout large regions. Research on and surveys of RTSV strains will require more attention in order to select RTSV-resistant varieties suitable for each region.

REFERENCES

ANJANEYULU, A. and JOHN, V.T. (1972) Strains of rice tungro virus. *Phytopathology* **62**: 1116–1119.

CABAUATAN, P.Q., CABUNAGAN, R.C. and KOGANEZAWA, H. (1994) Comparative transmission of two strains of rice tungro spherical virus in the Philippines. *International Rice Research Notes*, **19**(2): 10–11

CABAUATAN, P.Q., CABUNAGAN, C.R. and KOGANEZAWA, H. (1995) Biological variants of rice tungro viruses in the Philippines. *Phytopathology*, **85**: 77–81.

CABUNAGAN, R., FLORES, Z.M., COLOQUIO, E.C. and KOGANEZAWA, H. (1993a) Virus detection in varieties resistant to tungro (RTD). *International Rice Research Notes,* **18**(1): 22–23.

CABUNAGAN R. and KOGANEZAWA, H. (1993) Geographical distribution of resistant varieties to rice tungro disease (RTD). *International Rice Research Notes*, **18**(1): 21.

CABUNAGAN, R., FLORES, Z.M., HIBINO, H., MUIS, A., TALANCA, H., SUDJAK, S.M. and BASTIAN, A. (1989) Sporadic occurrence of tungro (RTV) in rice resistant to tungro spherical virus (RTSV). *International Rice Research Newsletter,* **14**: 13–14.

CABUNAGAN, R., FLORES, Z.M. and KOGANEZAWA, H. (1993b) Resistance to rice tungro spherical virus (RTSV) in rice germplasm. *International Rice Research Notes,* **18**: 21.

DAHAL, G., HIBINO H., CABUNAGAN, R., TIONGCO, E.R., FLORES, Z.M. and AGUIERO, V.M. (1990). Changes in cultivar reaction to tungro due to changes in 'virulence' of the leafhopper vector. *Phytopathology,* **80**: 659–665.

DAHAL, G., DASGUPTA, I., LEE, G. and HULL, R. (1992) Comparative transmission of, and varietal reaction to, three isolates of rice tungro virus disease. *Annals of Applied Biology,* **120**: 287–300.

HASANUDDIN, A. and HIBINO, H. (1989) Grain yield reduction, growth retardation, and virus concentration in rice plants infected with tungro-associated viruses. pp. 56–73. In: *Crop Losses due to Disease Outbreaks in the Tropics and Countermeasures*. Tropical Agriculture Research Series, No. 22. Japan: Tropical Agriculture Research Center (TARC).

HASANUDDIN, A., DAQUIOAG, R.D. and HIBINO, H. (1988) A method for scoring resistance to tungro (RTV). *International Rice Research Newsletter,* **13**: 13–14.

HEINRICHS, E.A. and RAPUSAS, H. (1983) Correlation of resistance to the green leafhopper, *Nephotettix virescens* (Homoptera: Cicadellidae) with tungro virus infection in rice varieties having different genes for resistance. *Environmental Entomology,* **12**: 201–205.

HIBINO, H. (1983) Transmission of two rice tungro-associated viruses and rice waika virus from doubly or singly infected source plants by leafhopper vectors. *Plant Disease,* **67**: 774–777.

HIBINO, H., TIONGCO, E.R., CABUNAGAN, R.C. and FLORES, Z.M. (1987) Resistance to rice tungro-associated viruses in rice under experimental and natural conditions. *Phytopathology,* **77**: 871–875.

HIBINO, H., DAQUIOAG, R.D., CABAUATAN, P.Q. and DAHAL, G. (1988) Resistance to rice tungro spherical virus in rice. *Plant Disease,* **72**: 843–847.

HIBINO, H., DAQUIOAG, R.D., MESINA, E.M. and AGUIERO, V.M. (1990) Resistances in rice to tungro-associated viruses. *Plant Disease,* **74**: 923–926.

IMBE, T. and HABIBUDDIN, B.H. (1989) *Studies on Breeding for Resistance to Tungro Disease of Rice in Malaysia.* Japan: Tropical Agriculture Research Center (TARC) and Malaysia: Malaysian Agricultural Research and Development Institute (MARDI).

INOUE, H. and RUAY-AREE, S. (1977) Bionomics of green rice leafhopper and epidemics of yellow orange leaf virus diseases in Thailand. pp. 117–121. In: Tropical Agriculture Research Series, No. 10. Japan: Tropical Agriculture Research Center (TARC).

INTERNATIONAL RICE RESEARCH INSTITUTE (1993) Varietal reactions in different countries. pp. 8–9. In: *Program Report for 1993.* Los Baños, Philippines: International Rice Research Institute.

KHUSH, G.S. (1989) Multiple disease and insect resistance for increased yield stability in rice. pp. 79–92. In: *Progress in Irrigated Rice Research.* Los Baños, Philippines: International Rice Research Institute.

LING, K.C. (1974) An improved mass screening method for testing the resistance of rice varieties to tungro disease in the greenhouse. *Philippine Phytopathology,* **10**: 19–30.

MANWAN, I., SAMA, A. and RIZVI, S.A. (1985) Use of varietal rotation in the management of tungro disease in Indonesia. *Indonesian Agriculture Research Development Journal,* **7**: 43–48.

RIVERA, C.T. and OU, S.H. (1967) Transmission studies of the two strains of rice tungro virus. *Plant Disease Reporter,* **51**: 877–881.

SAMA, S., HASANUDDIN, A., MANWAN, I., CABUNAGAN, R.C. and HIBINO, H. (1991) Integrated management of rice tungro disease in South Sulawesi, Indonesia. *Crop Protection,* **10**: 34–40.

SAXENA, R.C., MEDRANO, F.G. and BERNAL, C.C. (1991) Rapid screening for rice tungro resistance using viruliferous green leafhopper (GLH) nymphs. *International Rice Research Newsletter,* **16**: 10–11.

SHAHJAHAN, M., JALANI, B.S., ZAKRI, A.H., IMBE, T. and OTHMAN, O. (1990) Inheritance of tolerance to rice tungro bacilliform virus (RTBV) in rice (*Oryza sativa* L.). *Theoretical and Applied Genetics,* **80**: 513–517.

TAKAHASHI, Y., TIONGCO, E.R., CABAUATAN, P.Q., KOGANEZAWA, H., HIBINO, H. and OMURA, T. (1993) Detection of rice tungro bacilliform virus by polymerase chain reaction for assessing mild infection of plants and viruliferous vector leafhoppers. *Phytopathology,* **83**: 655–659.

Paper 8

New gene sources for tungro virus resistance in wild species of rice (*Oryza* spp.)

N. KOBAYASHI*, R. IKEDA†, D.A. VAUGHAN‡ and T. IMBE

International Rice Research Institute, PO Box 933, 1099 Manila, Philippines

INTRODUCTION

The use of resistant cultivars is one of the most efficient approaches to the management of disease. Host resistance to viral diseases can be classified into several types: resistance to viruses, tolerance of viruses and resistance to vectors (Fraser, 1990). Experience with rice tungro virus disease, referred to here as rice tungro disease, is that resistance to the main leafhopper vector has usually broken down within a few years of intensive cultivation (Hibino *et al.*, 1987) and the use of virus-resistant cultivars may be a more successful approach to control.

Strategies for breeding such cultivars need to be based on the particular characteristics of the virus(es) involved. Rice tungro disease is caused by a complex of two viruses, rice tungro bacilliform virus (RTBV) and rice tungro spherical virus (RTSV), and the most important vector is the green leafhopper (GLH) *Nephotettix virescens* Distant (Hibino, 1989). RTBV causes tungro symptoms and RTSV enhances these symptoms, although alone it causes no clear symptoms; RTSV plays an important role in virus transmission, as RTBV is not transmitted without acquisition of RTSV by GLH. Since the enzyme-linked immunosorbent assay (ELISA) was introduced (Bajet *et al.*, 1985), it has been possible to screen for resistance to each tungro virus separately. ELISA also makes it possible to distinguish between resistance to virus infection and to virus multiplication by measuring absorbance which is directly related to virus content.

At the International Rice Research Institute (IRRI), more than 40 000 accessions of cultivated rice (*Oryza sativa*) have been evaluated for resistance to each of the two tungro viruses and some cultivars have been found to be resistant (Hibino *et al.*, 1990; Koganezawa and Cabunagan, page 54). Some of these cultivars were resistant to RTSV infection and some are tolerant and/or resistant to multiplication of both RTBV and RTSV. Some commercial cultivars including IR26 were also found to be resistant to RTSV infection (Hibino *et al.*, 1988; Imbe *et al.*, 1993). Clear resistance to RTBV infection, however, has not yet been reported. The genetics of RTSV resistance has been studied and two genes were identified among several cultivars of diverse origin (Ikeda *et al.*, unpublished). Resistance to virus multiplication showed polygenic inheritance, which may be difficult to introduce into commercial cultivars.

Wild relatives of crops are an important reservoir of genes for disease resistance (Harlan, 1976). In rice, resistance to rice grassy stunt virus was transferred from the wild relative *O. nivara* to rice through backcross breeding (Khush, 1977). Breeding for other disease resistances from wild relatives of rice, such as resistance to blast caused by *Pyricularia grisea*, is also being carried out at IRRI. Screening of new and diverse gene sources for tungro resistance remains a major priority for rice breeders. Consequently, a comprehensive survey of tungro resistance was made in wild species of rice and African cultivated rice (*O. glaberrima*). Some of the results have already been reported (Kobayashi *et al.*, 1993a and b) and additional information is presented here.

TUNGRO SCREENING

Initial evaluation

To evaluate many germplasm accessions efficiently, 202 wild *Oryza* accessions, representing the genetic diversity and the range of distribution of all available species in the genus *Oryza*, were selected from the International Rice Germplasm Center (IRGC), IRRI. Classification of the *Oryza* species followed the taxonomy of Vaughan (1989).

---

*Present address: Faculty of Agriculture, Kobe University, Japan.
†Present address: National Agriculture Research Centre, Japan.
‡Present address: National Institute for Agrobiological Resources, Japan.

Table 1 Infection rates with rice tungro bacilliform virus (RTBV) and rice tungro spherical virus (RTSV) in *Oryza* species inoculated as seedlings

			Number of accessions											
				% infection with RTBV						% infection with RTSV				
Species	Genome	Tested	0	>0~≤20	>20~≤40	>40~≤60	>60~≤80	>80~≤100	0	>0~≤20	>20~≤40	>40~≤60	>60~≤80	>80~≤100
***O. sativa* complex**														
O. nivara	AA	52	0	0	0	1	8	43	4	8	11	16	8	5
O. rufipogon	AA	19	3	4	2	3	1	6	10	3	5	1	0	0
Natural hybrids	AA	35	0	0	1	2	8	24	6	12	6	7	4	0
O. glaberrima	AA	4	0	0	0	0	0	4	0	0	2	0	2	0
O. barthii	AA	7	0	0	0	0	2	5	1	5	1	0	0	0
O. meridionalis	AA	2	0	0	0	0	0	2	0	2	0	0	0	0
SUBTOTAL		119	3	4	3	6	19	84	21	30	25	24	14	5
***O. officinalis* complex**														
O. officinalis	CC	15	4	6	4	0	0	1	6	9	0	0	0	0
O. rhizomatis	CC	5	1	1	1	1	0	1	1	2	2	0	0	0
O. eichingeri	CC	5	0	2	2	0	0	1	2	2	0	1	0	0
O. malampuzhaensis	BBCC	3	0	0	1	0	1	1	1	2	0	0	0	0
O. minuta	BBCC	13	0	8	3	2	0	0	6	7	0	0	0	0
O. punctata	BB	4	0	0	1	1	2	0	2	2	0	0	0	0
O. punctata	BBCC	3	0	2	0	1	0	0	0	2	1	0	0	0
O. latifolia	CCDD	5	1	3	0	1	0	0	1	3	0	0	1	0
O. alta	CCDD	3	2	1	0	0	0	0	0	2	1	0	0	0
O. grandiglumis	CCDD	2	0	0	2	0	0	0	0	1	1	0	0	0
O. australiensis	EE	4	0	2	1	0	1	0	2	2	0	0	0	0
SUBTOTAL		62	8	25	15	6	4	4	21	34	5	1	1	0
***O. ridleyi* complex**														
O. longiglumis	Tetraploid	3	1	2	0	0	0	0	0	1	1	1	0	0
O. ridleyi	Tetraploid	5	2	2	0	1	0	0	0	2	1	1	0	1
SUBTOTAL		8	3	4	0	1	0	0	0	3	2	2	0	1
Not in any complex														
O. brachyantha	FF	5	1	4	0	0	0	0	5	0	0	0	0	0
TOTAL		194	15	37	18	13	23	88	47	67	32	27	15	6
Check varieties														
TN1 (susceptible)		1	0	0	0	0	0	1	0	0	0	1	0	0
Utri Merah (resistant)		1	0	0	1	0	0	0	0	1	0	0	0	0

*Following taxonomy of Vaughan (1989).

Of the 202 accessions, 194 were inoculated as seedlings grown in clay pots (13 cm diameter), using five seedlings per pot; on average, 34 plants (range 10 to 246) of each accession were tested. The seedlings were infested 14 days after sowing (DAS) using 10 viruliferous GLH adults per seedling for 4 h in mylar cages. Viruliferous GLHs were obtained by allowing adults of *N. virescens* to feed for 4 days on plants of Taichung Native 1 (TN1) which were doubly infected with RTBV and RTSV. Leaves were sampled individually 2–3 weeks after inoculation (WAI) and tested separately for RTBV and RTSV by ELISA. TN1 and Utri Merah (IRGC Acc. No. 16680) served as susceptible and resistant checks, respectively.

Eight accessions of *O. longistaminata* were evaluated by inoculating divided tillers of fully grown plants, since seeds were not available to provide seedlings. Fully grown plants of these eight accessions and two check cultivars, TN1 and Utri Merah, were cut 20 cm above the ground and divided into tillers. Each tiller was planted in a clay pot (13 cm diameter) and infested with 20 viruliferous GLH adults for 4 h. Leaves were sampled individually for ELISA at 2 WAI. Summarized results of the infections achieved are shown in Tables 1, 2 and 3.

***Oryza sativa* complex** Among wild species in the *Oryza sativa* complex, 10 out of 19 accessions of *O. rufipogon* were not infected with RTSV (Table 1) and three which originated from central Thailand (IRGC 105908, 105909 and 105910) were not infected with either virus (Table 2). Accessions of *O. rufipogon* from central Thailand showed a low overall incidence of infection. All 52 accessions of *O. nivara* were readily infected with RTBV, and only four were not infected with RTSV. All 35 accessions of natural hybrids between species having the AA genome were also infected with RTBV (Table 1). All accessions of *O. glaberrima*, *O. barthii* and *O. meridionalis* were infected with RTBV. *O. glaberrima* and a wild relative (*O. barthii)* showed systemic necrosis after tungro infection and few plants survived beyond 3 WAI. This systemic necrosis was the most severe reaction to tungro viruses observed and was not seen in other species.

***O. officinalis* complex** Compared with the *O. sativa* complex, the *O. officinalis* complex had more accessions which were resistant to tungro infection. Twenty-one of 62 accessions were not infected with RTSV and eight accessions were not infected with RTBV (Table 1). Two accessions of *O. officinalis* (IRGC 105100 and 105365) were not infected with either virus and the other two (IRGC 104672 and 105376) were only infected with RTSV (Table 2). Fourteen of the 15 accessions of the species showed low absorbance values in the ELISA test, indicating low concentration of each virus. One accession of *O. rhizomatis* (IRGC 103421) was not infected with either RTBV or RTSV, and one of *O. latifolia* (IRGC 105139) and two of *O. alta* (IRGC 100967 and 105685) were not infected with RTBV (Table 2). Most of the other accessions of the three species showed low infection rates and low absorbance in ELISA for RTBV and RTSV (Table 1). All accessions of *O. eichingeri, O. malampuzhaensis, O. minuta, O. punctata* and *O. australiensis* were infected with RTBV, although some accessions of these species showed a low infection rate (Table 1).

***O. ridleyi* complex** In the *O. ridleyi* complex, one accession of *O. longiglumis* (IRGC 105146) and two of *O. ridleyi* (IRGC 100821 and 101453) were not infected with RTBV (Table 2). Unlike most species of other complexes, all eight accessions of the *O. ridleyi* complex showed lower infection rates with RTBV than with RTSV. Five accessions of *O. brachyantha* were not infected with RTSV, and infection with RTBV was less than 20% (Table 1). One accession (IRGC 100115) of this species was not infected with either RTSV or RTBV (Table 2).

O. longistaminata All eight accessions of *O. longistaminata* tested were infected with RTBV, while two accessions were free of RTSV (Table 3).

In summary, of 202 accessions of 20 *Oryza* species (including *O. longistaminata)* and natural hybrids, 15 accessions of 8 species were not infected with RTBV (Table 1 and 3) and 51 accessions of 13 species and natural hybrids were not infected with RTSV. The 15 accessions with no RTBV infection may be suitable parents for use in RTBV resistance breeding and a selection of these was re-evaluated in further tests.

Table 2 Infection rates with rice tungro spherical virus (RTSV) in wild species that were not infected with rice tungro bacilliform virus (RTBV)

Species	IRGC Acc. No.	Origin	Plants tested (no.)	Infection rate (%) with RTSV
O. rufipogon	105908	Thailand	20	0
	105909	Thailand	23	0
	105910	Thailand	23	0
O. officinalis	104672	Malaysia	20	5
	105100	Brunei	28	0
	105365	Thailand	41	0
	105376	Thailand	29	3
O. rhizomatis	103421	Sri Lanka	18	0
O. latifolia	105139	Guatemala	25	4
O. alta	100967	Suriname	19	10
	105685	Brazil	29	11
O. longiglumis	105146	Indonesia	24	33
O. ridleyi	100821	Thailand	11	9
	101453	Malaysia	30	3
O. brachyantha	100115	Guinea	29	0

Table 3 Infection rates with rice tungro bacilliform virus (RTBV) and rice tungro spherical virus (RTSV) of cuttings of *O. longistaminata*

IRGC Acc. No.	Origin	Plants tested (no.)	Infection rate (%) with			
			RTBV	RTSV	RTBV+RTSV	None
101200	Nigeria	12	17	8	8	83
101207	Ivory Coast	15	7	0	0	93
103890	Senegal	15	60	67	40	13
103900	Tanzania	15	47	0	0	53
103902	Tanzania	15	40	27	14	47
103913	Tanzania	14	71	29	29	29
104151	Cameroon	15	47	20	7	40
105198	Ethiopia	15	80	20	20	20
Check varieties						
TN1 (susceptible)		15	67	100	67	0
Utri Merah (resistant)		15	27	33	7	47

Re-evaluation of accessions showing no infection with RTBV

Of 15 accessions not infected with RTBV, the 11 listed in Table 4 were re-evaluated using the same methods of inoculation and ELISA. Only one accession of *O. branchyantha* (IRGC 100115) and one of *O. ridleyi* (IRGC 101453) showed no RTBV infection, while the other nine accessions were infected with RTBV, albeit at low or moderate rates. Although RTBV infection rates in these accessions were lower than those in Utri Merah, absorbance values were higher.

Antibiosis to the vector *Nephotettix virescens*

Infection by tungro viruses can be influenced by GLH resistance (Habibuddin *et al.*, 1991). This can be attributed to the mechanism of GLH resistance which leads to decreased phloem feeding on resistant cultivars (Heinrichs and Rapusas, 1984; Karim and Saxena, 1991) and is associated with decreased rates of transmission of tungro viruses (Heinrichs and Rapusas, 1983; Dahal *et al.*, 1990). Therefore, evaluation for GLH resistance was carried out to confirm whether the low level of infection by RTBV in the 15 accessions of wild rice was due to resistance to virus or to GLH. GLH resistance was evaluated using the antibiosis test by determining survival of GLH nymphs on seedlings of the accessions (Kishino and Ando, 1978).

Table 4 Re-evaluation of the rice tungro bacilliform virus (RTBV)-resistant accessions of wild *Oryza* spp. using *Nephotettix virescens*

Species, cultivar	IRGC Acc. No.	Origin	Inoculated plants (no.)	Infection rate (%) with RTBV	RTSV	Absorbance of RTBV*
Experiment 1						
O. alta	100967	Suriname	32	9	22	1.73
	105685	Brazil	92	1	0	0.49
O. brachyantha	100115	Guinea	27	0	0	–
O. latifolia	105139	Guatemala	100	50	0	1.63
O. officinalis	104672	Malaysia	37	51	0	0.80
	105100	Brunei	41	22	0	0.50
TN1 (susceptible check)			50	100	100	1.79†
Utri Merah (resistant check)			49	76	4	0.10
Experiment 2						
O. rufipogon	105908	Thailand	35	40	14	0.54
	105909	Thailand	98	42	3	0.42
	105910	Thailand	27	37	0	0.27
O. ridleyi	100821	Unknown	91	4	7	0.18
	101453	Malaysia	38	0	0	–
TN1 (susceptible check)			50	98	96	2.17‡
Utri Merah (resistant check)			50	80	2	0.12

*Means of infected plants only.
†An absorbance value for one plant exceeded the range of the ELISA reader and a value of 3.0 was assumed.
‡Absorbance values of eight plants exceeded the range.

The tip of a newly expanded leaf for each accession was placed in a test tube (1.5 cm diameter × 15 cm long) with 2 ml water and infested with five 2nd-instar nymphs. The GLH colony used in the antibiosis test was that used for the virus inoculations. Each accession was replicated 10 times. Surviving nymphs were counted daily for three days after caging. Antibiosis was assessed from the weighted mean of nymph survival rate (%) calculated for each accession as follows (Ikeda, 1985):

$$\text{Weighted mean} = \frac{A_1 \times 1 + A_2 \times 2 \ldots + A_n \times n}{1 + 2 + 3 \ldots + n} \times 100$$

where $A_1, A_2 \ldots$ and A_n are nymph survival rates (%) 1, 2 … and n days after cageing, respectively.

The weighted mean of nymph survival rate on one accession of *O. rufipogon* (IRGC 105909) was not significantly different from that of the susceptible check TN1 (Table 5). Survival rates of the other two accessions of *O. rufipogon* were intermediate between TN1 and the vector-resistant check, IR64. GLH antibiosis of the other 12 accessions of wild species was as high as, or higher than, that of IR64.

These results suggest that the low or moderate infection of Acc. 105909 of *O. rufipogon* was due to RTBV resistance. It remains unknown whether resistance in other accessions is due to vector resistance and/or to virus resistance.

Evaluation using the alternative vector *N. nigropictus*

Eleven of the 12 accessions of wild rice which showed high antibiosis to *N. virescens* were re-evaluated for vector resistance and RTBV infection rate using *N. nigropictus*. For these experiments, females of *N. nigropictus* were collected on the IRRI farm and caged with plants of TN1 and *Echinochloa crus-galli* for oviposition. A 20-day-old seedling in a clay pot (7.5 cm diameter) was covered with a mylar cage and infested with 10 newly hatched nymphs. Numbers of surviving nymphs in each of ten cages were counted daily for 10 days after infestation. Five accessions which showed low antibiosis to *N. nigropictus* were evaluated for tungro infection with *N. nigropictus*. Thirty 14-day-old seedlings of each accession of *O. officinalis* were inoculated, following the procedures already described. Tillers of one accession of *O. rhyzomatis* were inoculated instead of seedlings.

Table 5   Nymph survival rate of *N. virescens* on detached leaves of wild species of rice (*Oryza* spp.)

	Species	IRGC Acc. No.	Nymph survival*(%)
Experiment 1	*O. rufipogon*	105908	49.7 cd
		105909	89.6 e
		105910	57.9 d
	O. officinalis	105365	5.6 a
		105376	29.8 bc
	O. rhizomatis	103421	17.3 ab
	O. ridleyi	101453	4.6 a
	TN1		94.3 e
	IR64		16.9 ab
Experiment 2	*O. officinalis*	104672	30.9 c
		105100	21.0 abc
	O. latifolia	105139	10.0 ab
	O. alta	100967	13.4 ab
		105685	24.9 bc
	O. longiglumis	105146	9.8 a
	O. ridleyi	100821	14.3 ab
	O. brachyantha	100115	10.5 ab
	TN1		86.9 e
	IR64		51.7 d

*Weighted means followed by a common letter are not significantly different at the 5% level by Duncan's multiple range test (DMRT) performed on the transformed data.

Three accessions of *O. officinalis* (IRGC 105100, 105365 and 105376) and one of *O. rhizomatis* (IRGC 103421) showed low antibiosis to *N. nigropictus* (Table 6). Nymph survival rates on another accession of *O. officinalis* (IRGC 104672) and one of *O. longiglumis* (IRGC 105146) were moderately or highly resistant; other accessions were highly resistant to *N. nigropictus*. Four accessions of *O. officinalis* were resistant to RTBV (Table 6), although the infection rate of the susceptible standard TN1 was low because of the inefficiency of *N. nigropictus* as vector. In contrast, the *O. rhizomatis* accession was infected with RTBV almost as readily as TN1.

Based on the antibiosis test and virus inoculations with the alternative vector *N. nigropictus*, the three accessions of *O. officinalis* (IRGC 105100, 105365 and 105376) were confirmed as resistant to RTBV infection. However, the resistance of IRGC 105100 was moderate in the re-evaluation tests (Table 4). Although a few other accessions showed no infection with RTBV in the re-evaluation with *N. virescens*, virus resistance could not be confirmed due to antibiosis to both vector species.

Necrotic reactions of *O. glaberrima* and *O. barthii* to tungro viruses

In the tungro screening, accessions of *O. glaberrima* and its close wild relative *O. barthii* showed severe systemic necrosis after inoculation, which had not been observed previously in any other *Oryza* species tested (Kobayashi and Ikeda, 1992). It was considered that the necrotic reaction of these two species might be a useful trait for the management of tungro disease, because infected plants would die quickly and would not be an inoculum source for further transmission. Such a reaction would be one of host hypersensitivity at the population level.

Therefore, 89 accessions of *O. glaberrima* and 53 of *O. barthii* were selected from the range of this material from different parts of Africa which are conserved in IRGC. Inoculation procedures followed those described previously. Twenty seedlings of each accession were inoculated and observed every three days after inoculation. At 30 days after inoculation, surviving plants were tested individually for RTBV and RTSV by ELISA; TN1 was the susceptible check.

Of the 89 accessions of *O. glaberrima*, 86 showed systemic necrosis (Table 7); in only one accession did all the seedlings die. Some seedlings of the remaining 85 accessions survived at least 30 days after inoculation. The survivors were either infected with RTBV alone or were not infected. Therefore, all plants infected with both viruses were assumed to show necrosis and die within 30 days of inoculation. Seedlings of one accession, IRGC 101438 from Guinea, showed either necrosis or the normal susceptible reaction in plants doubly infected with RTBV and RTSV. Three accessions, IRGC 101049 ('Africa'), 101855 (Burkina Faso), and 103616 (Mali) showed the typical symptoms found in *O. sativa*, such as yellow-orange discolouration of leaves but no necrosis, although they were infected with both RTBV and RTSV.

Table 6 Nymph survival rate of *N. nigropictus* and tungro infection by inoculation using *N. nigropictus*

Species	IRGC Acc. No.	Nymph survival*(%)	Inoculated plants (no.)	Infection rate (%) with	
				RTBV	RTSV
O. officinalis	104672	49.1 cd	28		
	105100	90.4 e	28	7	0
	105365	90.8 e	23	4	0
	105376	89.3 e	26	4	0
O. rhizomatis	103421	83.5 e	30	0	0
O. latifolia	105139	5.0 ab		30	3
O. alta	100967	13.1 b			
	105685	0.6 a			
O. longiglumis	105146	0.3 a			
O. ridleyi	100821	61.4 d			
	101453	39.0 c			
TN1		91.6 e	30	37	17
PTB2	6215	91.0 e			
ARC11554	21473	89.9 e			
Utri Merah	16680	–		30	0

*Weighted means followed by a common letter are not significantly different at the 5% level by Duncan's multiple range test (DMRT) performed on the transformed data.

Table 7 Distribution of systemic necrosis in *O. glaberrima* and *O. barthii* accessions from Africa caused by infection with rice tungro viruses

Origin	Number of accessions			
	O. glaberrima		*O. barthii*	
	Tested	With necrosis	Tested	With necrosis
West Africa				
Burkina Faso	8	7	1	1
The Gambia	1	1	2	2
Guinea	10	10	3	3
Côte d'Ivoire	5	5		
Liberia	14	14		
Mali	11	10	8	8
Nigeria	7	7	2	2
Senegal	12	12	3	3
Sierra Leone	2	2	2	2
Central West Africa				
Cameroon	4	4	5	5
Chad	11	11	16	15
Congo	1	1		
East Africa				
Sudan	1	1	5	4
Tanzania			2	2
South Africa				
Botswana			1	1
Zambia			1	1
'Africa'*	1	0	1	1
via CRRI†	1	1	1	1
TOTAL	89	86	53	51

*Country not specified.
†Central Rice Research Institute, India

66

The systemic necrosis was also observed in 51 of the 53 accessions of *O. barthii* (Table 7). In 27 of 51 necrosis-affected accessions, some surviving plants were infected with both RTBV and RTSV. Only two accessions, IRGC 100933 (Sudan) and 104112 (Chad), did not show any necrosis, although they were doubly infected.

Although the accessions tested in the study were collected from various African countries, most representatives of both species showed similar systemic necrosis. Therefore, necrosis is considered to be specific to the species. Since the geographic distribution of *O. glaberrima* and *O. barthii* is restricted to Africa, where neither tungro viruses nor their leafhopper vectors have been reported, such a severe reaction does not prejudice their survival in nature.

These species provide useful material for studying the mechanism of pathogenicity of tungro viruses and they are important as sources of hypersensitivity. Moreover, Cabauatan *et al.* (1993) suggested that *O. glaberrima* could be used as an indicator plant for RTSV.

DISCUSSION

From the screening of IRGC accessions, wild rice species were found to be sources of genes for tungro virus resistance. Although the apparent resistance to RTBV in many accessions could not be confirmed unequivocally due to their vector resistance, these accessions are possible candidates as resistance donors in future breeding programmes.

One accession of *O. rufipogon* (IRGC 105909) and three accessions of *O. officinalis* were identified as potential sources of RTBV resistance which is independent of vector resistance. New gene sources for RTSV resistance are important, because it is known that some RTSV-resistant cultivars of *O. sativa* are susceptible on Mindanao island, in the southern Philippines (Ebron *et al.*, personal communication). As *O. glaberrima* possesses a different type of resistance, it could be a source of tungro hypersensitivity genes of value in resistance breeding.

Accessions of *O. rufipogon* and *O. glaberrima* may be useful donors for breeding tungro-resistant cultivars, because the species have the same AA genome as *O. sativa* and their resistance can be incorporated by conventional breeding methods. For other species such as *O. officinalis*, significant advances have been made in overcoming hybrid sterilities between related species and *O. sativa* (Khush and Brar, 1988). Indeed, resistance genes to the brown planthopper (*Nilaparvata lugens*) and white-backed planthopper have already been incorporated into rice (Jena and Khush, 1990).

Breeding programmes are in progress to diversify gene sources of tungro resistance using these accessions of *Oryza* species.

ACKNOWLEDGEMENTS

These studies were carried out under the IRRI/Government of Japan Collaborative Project. The authors also thank the former IRRI Plant Pathologist, Dr H. Koganezawa, for providing the immunoglobulin for the ELISA.

REFERENCES

BAJET, N.B., DAQUIOAG, R.D. and HIBINO, H. (1985) Enzyme-linked immunosorbent assay to diagnose rice tungro. *Journal of Plant Protection in the Tropics*, **2**: 125–129.

CABAUATAN, P.K., KOBAYASHI, N., IKEDA, R. and KOGANEZAWA, H. (1993) *Oryza glaberrima*: an indicator plant for rice tungro spherical virus. *International Journal of Pest Management*, **39**: 273–276.

DAHAL, G., HIBINO, H., CABUNAGAN, R.C., TIONGCO, E.R., FLORES, Z.M. and AGUIERO, V.M. (1990) Changes in cultivar reactions to tungro due to changes in 'virulence' of the leafhopper vector. *Phytopathology*, **80**: 659–665.

FRASER, R.S.S. (1990). The genetics of resistance to plant viruses. *Annual Review of Phytopathology*, **28**: 179–200.

HABIBUDDIN, H., HADZIM, K., OTHMAN, O., IMBE, T. and OMURA, T. (1991) Selection of rice line Y1036 resistant to the green leafhopper and tungro disease. *MARDI Research Journal,* **19**: 169–175.

HARLAN, J.R. (1976) Genetic resources in wild relatives of crops. *Crop Science,* **16**: 329–333.

HEINRICHS, E.A. and RAPUSAS, H.R. (1983) Correlation of resistance to the green leafhopper, *Nephotettix virescens* (Homoptera: Cicadellidae) with tungro virus infection in rice varieties having different genes for resistance. *Environmental Entomology,* **12**: 201–205.

HEINRICHS, E.A. and RAPUSAS, H.R. (1984) Feeding, development, and tungro virus transmission by the green leafhopper, *Nephotettix virescens* (Distant) (Homoptera: Cicadellidae) after selection on resistant rice cultivars. *Environmental Entomology,* **13**: 1074–1078.

HIBINO, H. (1989) Insect-borne viruses of rice. *Advances in Disease Vector Research,* **6**: 209–241.

HIBINO, H., DAQUIOAG, R.D., CABAUATAN, P.Q. and DAHAL, G. (1988) Resistance to rice tungro spherical virus in rice. *Plant Disease,* **72**: 843–847.

HIBINO, H., MESINA, E.M. and AUIERO, V.M. (1990) Resistances in rice to tungro-associated viruses. *Plant Disease,* **74**: 923–926.

HIBINO, H., TIONGCO, E.R., CABUNAGAN, R.C. and FLORES, Z.M. (1987) Resistance to rice tungro-associated viruses in rice under experimental and natural conditions. *Phytopathology,* **77**: 871–875.

IKEDA, R. (1985) Studies on the inheritance of resistance to rice brown planthopper (*Nilaparvata lugens* Stål). and the breeding of resistant rice cultivars. *Bulletin of National Agriculture Research Center (Japan),* **3**: 1–54.

IMBE, T., HABIBUDDIN, H., IWASAKI, M. and OMURA, T. (1993) Resistance in some japonica rice cultivars to rice tungro spherical virus. *Japanese Journal of Breeding,* **43**: 549–556.

JENA, K.K. and KHUSH, G.S. (1990) Introgression of genes from *Oryza officinalis* Well *ex* Watt to cultivated rice, *O. sativa* L. *Theoretical and Applied Genetics,* **80**: 737–745.

KARIM, A.N.M.R. and SAXENA, R.C. (1991) Feeding behavior of three *Nephotettix* species (Homoptera: Cicadellidae) on selected resistant and susceptible rice cultivars, wild rice, and graminaceous weeds. *Journal of Economic Entomology,* **84**: 1208–1215.

KHUSH, G.S. (1977) Disease and insect resistance in rice. *Advances in Agronomy,* **29**: 265–341.

KHUSH, G.S. and BRAR, D.S. (1988) Wide hybridization in plant breeding. pp. 141–188. In: *Plant Breeding and Genetic Engineering.* ZAKRI, A.H. (ed.). Malaysia: SABRAO.

KISHINO, K. and ANDO, Y. (1978) Insect resistance of the rice to the green rice leafhopper *Nephotettix cincticeps* Uhler. 1. Laboratory technique for testing the antibiosis. *Japanese Journal of Applied Entomology and Zoology,* **22**: 166–177.

KOBAYASHI, N. and IKEDA, R. (1992) Necrosis caused by rice tungro viruses in *Oryza glaberrima* and *O. barthii.* *Japanese Journal of Breeding,* **42**: 885–890.

KOBAYASHI, N., IKEDA, R. and VAUGHAN, D.A. (1993a) Resistance to rice tungro viruses in wild species of rice (*Oryza* spp.). *Japanese Journal of Breeding,* **43**: 247–255.

KOBAYASHI, N., IKEDA, R., DOMINGO, I.T. and VAUGHAN, D.A. (1993b) Resistance to infection of rice tungro viruses and vector resistance in wild species of rice (*Oryza* spp.). *Japanese Journal of Breeding,* **43**: 377–387.

VAUGHAN, D.A. (1989) *The Genus Oryza L. Current Status of Taxonomy.* IRRI Research Paper Series 138. Los Baños, Philippines: International Rice Research Institute.

Present status of rice tungro disease in India

A.K. CHOWDHURY

Department of Plant Pathology, Bidhan Chandra Krishi Viswavidyalaya, PO Krishi Viswavidyalaya, Mohanpur, West Bengal 741252, India

INTRODUCTION

Rice tungro virus disease, subsequently referred to as rice tungro disease (RTD), is one of the most important diseases of rice in India where the crop is a staple food that provides at least 30% of all calories consumed by the population. The disease was first identified in India from the state of West Bengal (Raychaudhuri and Ghosh, 1967; Raychaudhuri *et al.*, 1969) soon after it was first described and transmitted experimentally by leafhopper vectors in the Philippines (Rivera and Ou, 1965). These developments occurred during a period of great change in agricultural practices in India and elsewhere in the region with the introduction of high-yielding crop varieties, efficient use of inorganic nitrogenous fertilizers and other innovations.

Two locally bred rice varieties, namely 'Padma' and 'Jaya', and a number of the International Rice Research Institute (IRRI) 'IR' varieties were released in India in 1969. The good performance of these semi-dwarf, non-photosensitive rice varieties led to their increased adoption. After a few years of intensive cultivation all such varieties succumbed during epidemics of RTD in many states of north-eastern and south-eastern India (John, 1970). RTD epidemics later occurred sporadically in north-eastern India, particularly in West Bengal, Assam, Tripura, Manipur and in Kerala in 1973–74 (Anjaneyulu and Chakrabarti, 1977). Since 1975 the prevalence of the disease has increased in the southern states where rice is grown extensively throughout the year. In the 1984–85 cropping seasons RTD was widespread and attacked 80 000 ha planted mostly with high-yielding semi-dwarf varieties. A sudden severe outbreak of the disease during 1990 in considerable areas of West Bengal stimulated scientists, rice growers and government administrators to pay increased attention to the complex ecology of RTD in relation to its vectors, hosts and the environment, so as to formulate a strategy for minimizing the losses sustained. Since RTD became established, several institutions in India initiated studies on various aspects of the disease and this report summarizes the current status of this research.

RICE PRODUCTION

Rice is an indigenous crop of India that is cultivated two or three times a year depending on rainfall and the availability of irrigation water. The country can be divided into six zones based on rainfall and other climatological parameters: arid, semi-arid, dry sub-humid, moist sub-humid, humid and super-humid. Rice is grown in all six zones, mostly under rainfed conditions and is known as the *kharif* crop. More than 80% of total *kharif* plantings are of tall Indica-type varieties and the seasons range from May–June to November–December. The second *rabi* rice crop (December–May) is generally transplanted in the winter months, mostly with non-photosensitive dwarf or semi-dwarf high-yielding varieties. In some zones a third crop known as *aus* (May–October) is cultivated which provides a 'green bridge' between the *kharif* and *rabi* crops.

RTD is found mostly in sub-humid areas; it does not appear in arid or super sub-humid zones due to the almost complete absence of the principal leafhopper vectors including *Nephotettix virescens* (Distant) and *N. nigropictus* (Stål). Rice green leafhoppers (GLH) in India are seasonal and found mainly during the mid-growth stage of *kharif* rice. The prevalence of GLH is largely governed by temperature, humidity and rainfall, but also by the susceptibility to infestation of the varieties grown. Under Indian conditions many of the tall Indica varieties are tolerant to RTD, yet outbreaks of the disease have occurred in the past which affected many such varieties.

The status of RTD in India has been reviewed by Mishra (1977) and Mukhopadhyay (1984, 1986); the major outbreaks of RTD and GLH appear sporadically and only in some provinces (Table 1). RTD is currently widespread in India: it continues to occur erratically for reasons not yet clearly understood.

TUNGRO RESEARCH

The implementation of research findings leading to the use of vector-resistant varieties, insecticides and improved cultural practices, not only in India but throughout South and South-East Asia, has contributed to the control of RTD. Nevertheless, the disease remains a major problem of rice farmers in the region and one that is highly unpredictable.

Research on RTD in India began once the disease was first attributed to a virus in 1968; there has been a major thrust to understand its epidemiology, and on vector ecology and disease management by operational control or breeding disease/vector-resistant varieties (John, 1968; Mukhopadhyay, 1984; Mukhopadhyay *et al.*, 1985, 1986; Saxena and Anjaneyulu, 1987). The Directorate of Rice Research, regional rice research stations and several agricultural universities are presently involved in research on RTD in India.

Virus epidemiology

Rice tungro viruses seem better adapted to rice than to weed or wild hosts. Studies of potential weed hosts in different agro-climatic regions of India have suggested several graminaceous weed species as potential sources of tungro viruses, but the results are somewhat contradictory (Raychaudhuri and Ghosh, 1967; Mishra *et al.*, 1973; Rao and Anjaneyulu, 1978; Tarafder and Mukhopadhyay, 1980). Weeds commonly found in rice fields and in the off-seasons which are potential reservoirs of tungro viruses include *Echinochloa colonum, Eleusine indica, Hemarthria compressa* and *Polypogon monspeliensis.*

E. indica, E. colonum and *Pennisetum typhoides,* which have been reported as weed hosts (Prasada Rao and John, 1974), were not infected, even after repeated inoculation with a local virus isolate at the Central Rice Research Station in Cuttack, whereas *E. indica* and *E. colonum* were susceptible in Kalyani, West Bengal (Tarafder and Mukhopadhyay, 1980). Moreover, *E. indica, E. colonum, Eragrostis tenella, Leersia hexandra* and *Sorghum bicolor* from the Philippines (Rivera *et al.*, 1969; Wathanakul, 1964); *E. colonum* and *E. indica* from Malaysia (Ting, 1971) and *E. colonum, L. hexandra* and *Leptochloa chinensis* from Thailand (Hino *et al.*, 1974) did not give any positive reaction when inoculated with the Cuttack isolate. In all cases, detection of virus in the weeds was by visual symptoms or by back inoculations to rice using vectors and not by serology, which may be one of the reasons for the contradictory results.

Table 1 Major outbreaks of rice tungro disease and green leafhopper in India

Year	Rice tungro disease	Green leafhopper
1969	West Bengal; Bihar; Uttar Pradesh	–
1981	West Bengal; Bihar, Madhya Pradesh	West Bengal; Bihar; Madhya Pradesh
1982	–	West Bengal; Orissa; Madhya Pradesh
1983	–	West Bengal; Orissa; Madhya Pradesh
1984	Tamil Nadu; Andhra Pradesh	Tamil Nadu; Andhra Pradesh
1985	–	Bihar
1990	West Bengal	–
1994	West Bengal	–

Source: Mishra, 1977; Mukopadhyay, 1984, 1986.

A few wild rices can also harbour the tungro viruses when inoculated by infective vectors: Rao and Anjaneyulu (1978) and Anjaneyulu *et al.* (1982) inoculated 39 species of plants, including several wild rices (*Oryza nivara, O. perennis, O. barthii, O. australiensis, O. brachyantha, O. cichingeri* and *O. punctata*) which developed RTD-like symptoms. However, a critical role of these weeds and wild rices in the perennation of RTD has not been demonstrated. A marked variation was observed in the symptoms in different wild rice species, some of which may be symptomless carriers.

It is known that overlapping rice crops and rice stubbles that survive between seasons play a significant role in the perennation of RTD. Under Indian conditions stubbles often remain in the field for a considerable time depending on the intensity of cropping — almost six months in rainfed or mono-cropped areas until the land is prepared for the following *kharif* crop. Such stubbles regenerate as ratoons if adequate soil moisture is available and then become sources of infection if the previous crop was infected. Stubbles, ratoons and volunteer rice not only act as reservoirs of the virus, but also

70

facilitate the survival of GLH during the off-season (Chakrabarty *et al.*, 1985). The retention of virus and the efficiency of stubbles as reservoirs of the virus varies with rice variety, age and other conditions. Tarafder and Mukhopadhyay (1979, 1980) determined the longevity of virus in stubbles of different crop seasons using both high-yielding and local varieties. A variety having high susceptibility to virus both in field and laboratory conditions was generally a good source of virus or its strains (Mukhopadhyay and Chowdhury, 1973; Anjaneyulu and John, 1972; Mishra *et al.*, 1976; Chowdhury, 1993).

Vector ecology

In India *N. virescens* and *N. nigropictus* are the principal vectors of RTD. Populations of GLH are seasonal and depend mainly on meteorological conditions, particularly temperature, rainfall and relative humidity. Populations of GLH in all agro-climatic zones are usually highest during September–November (*kharif* season), but the date of the peak population differs between years for reasons not as yet determined. One of the causes of the sporadic incidence of RTD in India may be the irregular and weather-dependent incidence of vectors, together with differences in the availability of sources of virus inoculum. Peak populations of GLH in West Bengal occur during September–October, when the temperature ranges between 30°C and 35°C and relative humidity is 60–70%. Mukhopadhyay and Mukhopadhyay (1987) established a correlation between peak light-trap catches and rainfall. Peak catches usually occurred 60±10 days after peak monsoon rains, the lag period varying between years and locations. *N. virescens* populations in rice were greater than those of *N. nigropictus*. Through laboratory studies it was established that meteorological conditions influence the biology of *N. virescens* (Chakravarti *et al.*, 1979). Decreasing temperature increased the duration of the life cycle, including both the pre-ovipositional and nymphal periods. The life cycle was also extended at low relative humidity.

The duration of the life cycle of GLH is governed by environmental conditions, and threshold values for the maximum and minimum temperatures for its frequency distribution are 30°C and 15°C, respectively; GLH does not multiply below the minimum threshold temperature. Large fluctuations in the incidence of GLH are observed with differences of temperature in different seasons of different regions. In tropical areas there is little variation between summer and winter temperatures — rice crops remain vulnerable to RTD irrespective of season and GLH vectors occur throughout the year. In the subtropics temperatures fall below 15°C during the winter and GLH normally survive as nymphs, which have a prolonged life span; spread of RTD is limited because GLH populations are low. In the other agro-climatological regions of India, RTD epidemics occur mostly in the *kharif* season or post-monsoon months, especially in the north-east region. Thus RTD outbreaks occur in climatologically 'high-risk' areas when infective GLH populations exceed a critical threshold.

Host/virus/vector interactions

Damaging outbreaks of RTD are the outcome of complex interactions involving hosts, viruses and vectors. Studies at a number of institutes in India show many similarities with results obtained elsewhere in South-East Asia. John (1968) first identified and characterized RTD by vector transmission studies and Mukhopadhyay and Chowdhury (1973) later made detailed studies on the epidemiology of the disease. They established a non-persistent relationship of the virus with the vector and wide variation in rates of virus acquisition from and inoculation to different rice varieties. GLH survived better on aged seedlings of susceptible varieties than on young seedlings and they preferred to move from younger to older seedlings, irrespective of variety (Mukhopadhyay and Chattopadhyay, 1975). Age of the host plants has a profound effect on the spread of RTD. The results of tests on the susceptibility of a number of rice varieties, including tall Indica and high-yielding varieties bred in India and others from IRRI, to both tungro viruses and GLH, using the Indian isolates of virus, were similar to those found by other workers elsewhere.

MANAGEMENT OF TUNGRO DISEASE

In India the use of resistant varieties and chemical control of GLH have received much attention in attempts to control RTD. However, cultural practices recommended on the basis of regional research are also practised by Indian farmers. Attempts to demonstrate a relationship between meteorological conditions and the peak appearance of GLH with a view to forecasting suggested the possibility of using data on the monsoon rains to predict subsequent vector populations (Mukhopadhyay and Mukhopadhyay, 1987; Mukhopadhyay *et al.*, 1988; Mukhopadhyay and Mukhopadhyay, 1992). Recommended measures for RTD control as practised by Indian farmers are summarized below.

Chemical control

Any chemical control recommendations should be technically and economically justifiable. Moreover, insecticides should be the method of last resort to be used only when other measures are likely to be ineffective in restricting the disease below economic threshold levels. Chemical control of GLH has been studied intensively in India (Rao and Anjaneyulu, 1979; Shukla and Anjaneyulu, 1980; Satapathy and Anjaneyulu, 1986; Mukhopadhyay *et al.*, 1986). The field application of insecticides generally depends on the time of appearance of vectors, the resistance of the host variety towards GLH and virus, and also on field conditions. Of the insecticides used to control GLH, granular formulations and those having a wide range of pesticidal activity are generally preferred by farmers. However, synthetic pyrethroids have also been used to control RTD and its vector (Satapathy and Anjaneyulu, 1984). The general tendency is for Indian farmers to use more insecticides in the *boro* season than in the rainfed *kharif* crop.

The needs-based application of pesticides is a general strategy in plant protection, but prophylactic applications of pesticides are justified to prevent the entry of GLH from nearby fields in areas prone to RTD, or where an overlapping crop sequence facilitates the perpetuation of the disease. As mentioned above, RTD incidence is greatest late in the *kharif* season due to high GLH populations; from these crops it is spread to the seed-bed of *boro* rice by migrating GLH, particularly in north-eastern India. A large area of this region grows *boro* rice mostly with high-yielding varieties and RTD occurs at a low incidence in some areas. Bidhan Chandra Krishi Viswavidyalaya (BCKV) experience suggests that proper management of seed-beds to minimize the spread of RTD is one of the important components for tungro management and justifies one or two applications of a systemic insecticide to protect the seedlings from infection.

The amount of spread that occurs depends on the vector population density and the number and potency of virus sources present inside and outside the fields at risk. In India, tungro epidemics are noted mainly in *kharif* (rainfed) rice, when there is no other means to check spread other than the use of insecticides. Recommended schedules to control GLH and other rice pests have been developed regionally from field and laboratory evaluations (Rajak, 1986). An operational project conducted in West Bengal indicated that two seed-bed applications and one field application of a systemic insecticide to susceptible varieties, and one seed-bed and one subsequent application to tolerant varieties are required to reduce RTD infection (Chakrabarty, 1985).

Resistant/tolerant varieties

The best and cheapest means of controlling any disease is by growing resistant varieties (Mohanty *et al.*, 1989). Both high-yielding and local varieties are widely planted in the *kharif* season, while high-yielding IR varieties or cultivars bred in national programmes are used in the *boro*. Many of these varieties were at first resistant to RTD but, after a few years of intensive cultivation, they became susceptible and it was necessary for Indian scientists to breed new, resistant rice varieties at regular intervals. In the screening tests over many years at BCKV, particularly in respect to cv. Jaya and several other varieties, a gradual loss of resistance became apparent. Thus any particular resistant variety remains effective for only a limited period.

Adjustment of planting date

Adjustment of date of planting influences the incidence of tungro disease but variations may occur between years due to fluctuations in the incidence of GLH. Early planting of rice, especially in north-east India, is generally recommended to escape RTD. Where late planting is unavoidable, it is advisable to grow resistant/tolerant varieties and use one to two foliar sprays of insecticide. Field studies established that variation in RTD incidence was governed mainly by vector populations and the availability of virus inoculum (Chakrabarty, 1985; Shukla and Anjaneyulu, 1981). Disease incidence is expected to be lowest in plantings made in January (*rabi* crop) and July (*kharif* crop), especially in the eastern Indian provinces. Past records showed that late planting in the *kharif* season during 1969 (John, 1970) and 1973 (Anjaneyulu and Chakrabarti, 1977) was followed by severe damage to several thousand hectares of rice.

Cultural practices

There has been no economic evaluation of the reduction in incidence of RTD due to the adoption of the cultural practices recommended in India. Those commonly advised include:

- removal of infected stubbles/weeds by ploughing immediately after harvest
- removal and destruction of RTD-affected plants within crops, especially during the initial stage of infection, followed by appropriate application of an insecticide

- avoiding as far as practicable the overlapping of successive rice crops
- adoption of non-host crops in RTD-endemic areas
- use of nitrogenous fertilizer which is sometimes useful in compensating for the losses due to the disease when infection occurs late.

Integrated approach

Integrated management practices have been worked out and the findings are communicated through group meetings, mass communication media, field demonstrations and on-farm trials.

CONCLUSIONS

RTD in India was first observed in 1967 and systematic studies on the disease were initiated as soon as the viral nature of the causal pathogen was inferred. Considerable progress has been achieved in understanding the ecology of the vector and the epidemiology of the disease, sources of resistance, the nature of the causal viruses, and vector prediction through meteorological relationships and other means. Nevertheless, many questions need answering in order to safeguard rice, as the most important staple cereal, from what is a very unpredictable disease, not only in India but also in other countries of South and South-East Asia. Indian farmers encounter several constraints in adopting control measures, i.e. the unpredictable incidence of GLH vectors and disease; lack of rapid methods of virus identification; inadequate technology transfer; and socio-economic constraints.

The national agricultural network system involving rice research institutes, agricultural universities and other agriculture-orientated research centres has, through individual or collaborative efforts, focused on various aspects of RTD. On the basis of those observations the following priorities can be set for future research:

- know the virus(es) and their relationships with hosts and vectors
- search for durable sources of resistance to both viruses and vectors within areas where rice is indigenous
- carry out quantitative epidemiology studies on forecasting of viruses and vectors, including assessments of the possibility of migration of GLH between regions or countries
- integration of management options.

Success in minimizing the incidence of RTD — and the losses caused — requires close collaboration with other countries for joint research projects with adequate funding, and exchange of scientific ideas and technologies through seminars, workshops and advanced training on the whole range of new technologies necessary for virus detection and overall crop improvement.

REFERENCES

ANJANEYULU, A. and CHAKRABARTI, N.K. (1977) Geographical distribution of rice tungro virus disease and its vector in India. *International Rice Research Newsletter,* **2**: 15–16.

ANJANEYULU, A. and JOHN, V.T. (1972) Strains of rice tungro virus. *Phytopathology,* **62**: 1116–1119.

ANJANEYULU, A., SHUKLA, V.D., RAO, G.M. and SINGH, S.K. (1982) Experimental host range of rice tungro virus and its vector. *Plant Disease,* **66**: 54–56.

CHAKRABARTY, S.K. (1985) *Operational Techniques for Controlling Rice Tungro Virus Disease in West Bengal.* PhD Thesis, West Bengal, India: Bidhan Chandra Krishi Viswavidyalaya.

CHAKRABARTY, S.K., NATH, P.S., CHOWDHURY, A.K. and MUKHOPADHYAY, S. (1985) Studies on the off-season incidence of rice green leafhoppers. pp. 87–90. In: *Use of Traps for Pest/ Vector Research and Control.* MUKHOPADHYAY, S. and GHOSH, M.R. (eds). West Bengal, India: Bidhan Chandra Krishi Viswavidyalaya.

CHAKRAVARTI, S., GHOSH, A.B. and MUKHOPADHYAY, S. (1979) Biology of the green leafhopper *Nephotettix virescens. International Rice Research Newsletter,* **4**: 16–17.

CHOWDHURY, A.K. (1993) *Final Report of Research Scheme on Monitoring of Rice Tungro Virus and its Vectors in West Bengal.* Government of West Bengal Project. West Bengal, India: Bidhan Chandra Krishi Viswavidyalaya.

HINO, T., WATHANAKUL, L., NABBEERONG, N., SURIN, P., CHAIMONGKOL, U., DISTHAPORN, S., PUTTA, M., KERDCHOKCHAI, D. and SURIN, A. (1974) Studies on yellow orange leaf disease in Thailand. pp. 1–67. *Technical Bulletin* No. 7. Japan: Tropical Agricultural Research Center. (TARC).

JOHN, V.T. (1968) Identification and characterization of tungro virus disease of rice in India. *Plant Disease Reporter,* **52**: 871–875.

JOHN, V.T. (1970) Yellowing disease of paddy. *Indian Farming*, **20**: 27–30.

MISHRA, M.D. (1977) Investigation on rice tungro virus in India. pp. 109–116. In: *Tropical Agriculture Research Series* No. 10. Japan: Tropical Agriculture Research Center (TARC).

MISHRA, M.D., GHOSH, A., NIAZI, F.R., BASU, A.N. and RAYCHAUDHURI, S.P (1973) The role of graminaceous weeds in the perpetuation of rice tungro virus. *Journal of Indian Botanical Society,* **52**: 176–183.

MISHRA, M.D., NIAZI, F.R., BASU, A.N., GHOSH, A. and RAYCHAUDHURI, S.P. (1976) Detection and characterization of a new strain of rice tungro virus in India. *Plant Disease Reporter,* **60**: 23–25.

MOHANTY, S.K., BHAKTVATSALAM, G. and ANJANEYULU, A. (1989) Identification of field resistant rice cultivars for tungro disease. *Tropical Pest Management,* **35**: 48–50.

MUKHOPADHYAY, S. (1984) Ecology of rice tungro virus and its vectors. pp. 139–164. In: *Virus Ecology.* MISHRA, A. and POLASA, H. (eds). New Delhi, India: South Asian Publishers.

MUKHOPADHYAY, S. (1986) Virus diseases of rice in India. pp. 111–122. In: *Vistas in Plant Pathology.* VARMA, A. and VERMA, J.P. (eds). India: Malhotra Publishing House.

MUKHOPADHYAY, S. and CHATTOPADHYAY, K. (1975) Preferential feeding of green leafhopper. *International Rice Commission Newsletter,* **24**: 76–80.

MUKHOPADHYAY, S. and CHOWDHURY, A.K. (1973) Some epidemiological aspects of tungro virus disease of rice in West Bengal. *International Rice Commission Newsletter,* **19**: 9–12.

MUKHOPADHYAY, S. and MUKHOPADHYAY, S. (1987) Lag correlation between the peak monsoon rains and peak appearance of rice green leafhoppers in West Bengal. *Proceedings of Indian National Science Academy,* **B53**: 189–191.

MUKHOPADHYAY, S. and MUKHOPADHYAY, S. (1992) On forecasting the peak trap catches of *Nephotettix* spp. (*N. virescens* Distant and *N. nigropictus* Stål) vectors of rice tungro virus in West Bengal. *International Journal of Tropical Plant Diseases,* **10**: 37–41.

MUKHOPDHYAY, S., CHAKRAVARTI, S. and MUKHOPADHYAY, S. (1985) The use of light traps for studying the biometeorological relation of the rice green leafhopper. pp. 91–102. In: *Use of Traps for Pest/Vector Research and Control.* MUKHOPADHYAY, S. and GHOSH, M.R. (eds). West Bengal, India: Bidhan Chandra Krishi Viswavidyalaya.

MUKHOPADHYAY, S., MUKHOPADHYAY, S., SARKAR, T.K., SARKAR, S. and NATH, P.S. (1986) Ecology of rice green leafhoppers, the vectors of rice tungro virus in West Bengal. pp 32–51. In: *Ricehoppers, Hopper-borne Viruses and their Integrated Management.* MUKHOPADHYAY, S. and GHOSH, M.R. (eds). West Bengal, India: Bidhan Chandra Krishi Viswavidyalaya.

MUKHOPADHYAY, S., MUKHOPADHYAY, S. and SARKAR, T.K. (1988) Effect of some meteorological factors on the incidence of rice green leafhoppers. pp. 189–197. In: *Agrometeorological Information for Planning and Operation in Agriculture with particular Reference to Plant Protection.* KRISHNAMURTHY, V. and MATHYS, G. (eds). Geneva.

PRASADA RAO, R.D.V.J. and JOHN, V.T. (1974) Alternate host of rice tungro virus and its vector. *Plant Disease Reporter,* **58**: 856–860.

RAJAK, R.L. (1986) Rice hoppers and their integrated management. pp. 86–94. In: *Ricehoppers, Hopper-borne Viruses and their Integrated Management.* MUKHOPADHYAY, S. and GHOSH, M.R. (eds). West Bengal, India: Bidhan Chandra Krishi Viswavidyalaya.

RAO, G.M. and ANJANEYULU, A. (1978) Host range of rice tungro virus. *Plant Disease Reporter,* **62**: 955–957.

RAO, G.M. and ANJANEYULU, A. (1979) Carbofuran prevents rice tungro virus infection. *Current Science,* **48**: 116–117.

RAYCHAUDHURI, S.P. and GHOSH, A. (1967). Occurrence of paddy virus and virus-like symptoms in India. pp. 59–65. In: *The Virus Diseases of Rice Plants.* Baltimore, Maryland: John Hopkins Press.

RAYCHAUDHURI, S.P., CHENULU, V.V. and GANGULY, B. (1969) Electron microscopic studies on rice and wheat viruses in India. pp. 25–26. In: *Electron Microscopy in Life Science.* Calcutta, India.

RIVERA, C.T. and OU, S.H. (1965) Leafhopper transmission of 'tungro' disease of rice. *Plant Disease Reporter,* **49**: 127–135.

RIVERA, C.T., LING, K.C. and OU, S.H. (1969) Suspect host range of rice tungro virus. *Philippine Phytopathology,* **5**: 16–17.

SATAPATHY, M.K. and ANJANEYULU, A. (1984) Use of cypermethrin, a synthetic pyrethroid in the control of rice tungro virus disease and its vector. *Tropical Pest Management,* **30**: 170–178.

SATAPATHY, M.K. and ANJANEYULU, A. (1986) Prevention of rice tungro virus disease and control of the vector with granular insecticides. *Annals of Applied Biology,* **108**: 503–510.

SAXENA, R.C. and ANJANEYULU, A. (1987) Tungro situation in India. pp. 51–53. In: *Proceedings of the Workshop on Rice Tungro Virus.* Ministry of Agriculture, Indonesia: AARD-Maros Research Institute for Food Crops.

SHUKLA, V.D. and ANJANEYULU, A. (1980) Evaluation of systemic insecticides for control of rice tungro. *Plant Disease,* **64**: 790–792.

SHUKLA, V.D. and ANJANEYULU, A. (1981) Adjustment of planting date to reduce tungro disease. *Plant Disease,* **65**: 409–411.

TARAFDER, P. and MUKHOPADHYAY, S. (1979) Potential of rice stubble in spreading tungro in West Bengal, India. *International Rice Research Newsletter,* **4**: 18.

TARAFDER, P. and MUKHOPADHYAY, S. (1980) Further studies on the potential of weeds to spread tungro in West Bengal, India. *International Rice Research Newsletter,* **5**: 10.

TING, W.P. (1971) Studies on penyakit merah disease of rice. II. Host range of the virus. *Malaysian Agricultural Journal,* **48**: 10–12.

WATHANAKUL, L. (1964) *A Study on the Host Range of Tungro and Orange Leaf Viruses of Rice.* MSc Thesis, University of Philippines, College of Agriculture.

Status of rice tungro disease in the Philippines: a guide to current and future research

A.R. BARIA

Philippine Rice Research Institute, Maligaya, Munoz, 3119 Nueva Ecija, Philippines

INTRODUCTION

Rice is considered to be the 'bread of life' of the Filipino people and although production is increasing it has not consistently met demand. The annual increases in production of 1.3% in recent years are less than the population growth rate of 2.5%. Moreover, the available land area is limited (3.2 million ha); half of this area is unsuitable for rice because it is either acidic or saline, or prone to drought or flood. Sustained rice production is also limited by various factors. One of these is rice tungro virus disease (RTVD), or rice tungro disease (RTD). This disease is caused by a complex of two viruses — rice tungro spherical virus (RTSV) and rice tungro bacilliform virus (RTBV) — that are transmitted semi-persistently by green leafhoppers (GLH) of which *Nephotettix virescens* is the most important. RTD has been, and remains, a continuing problem in so-called 'hot spot' areas of the country and also appears sporadically elsewhere.

Due to the lack of more effective alternative control measures, or simply because of ignorance, farmers have frequently resorted to the use of insecticides to control RTD in recent decades. Better alternative management practices are required so that farmers can overcome the tungro problem. Integrated pest management (IPM) research and management extension aspects should focus on the categorization of control options and prioritization of the importance of control measures relative to the overall sustainability of the agro-ecosystem.

This paper considers the occurrence of RTD in the Philippines and some contributory factors; recommended control measures, and the extent to which they are adopted, are also considered. Farmers' control practices and options at the community level, assessed from interviews conducted in Bukidnon Province, are discusssed and the current thrust of the national research and development (R&D) network is described. Some ongoing research activities are also listed, with proposals and recommendations for possible collaboration with various institutions.

TUNGRO DISEASE PROFILE: GEOGRAPHICAL DISTRIBUTION

In recent years tungro has continued to be considered the most important disease of rice in the Philippines. It is widely distributed in the country, particularly in the Cagayan Valley, and in the Bicol and Mindanao Regions (Figure 1), although data from detailed surveys are not widely available. Sporadic outbreaks occur mainly in regions where planting continues throughout the year and leads to overlapping rice crops at different growth stages. Moreover, many previously unreported outbreaks of RTD were mentioned recently during workshops conducted by the Philippine Rice Research Institute (PhilRice), in in-country surveys of rice areas and also in reports from network members. This further supports the claim of country scientists that reports of RTD have underestimated disease incidence in the last 10 years or so, implying the need to develop further the surveillance system for tungro, especially in key rice-growing areas.

Where rice is planted throughout the year, RTD occasionally causes epidemics and damages the crop over a huge area. This may be partly attributed to the widespread cultivation of rice varieties with similar genetic background. The prevalence of tungro in the Philippines in the early 1970s was associated with the introduction of new, high-yielding varieties, some of which were susceptible. High-yielding varieties are now planted in almost 80% of the rice growing regions of the country.

The relative susceptibility of these varieties, abundant sources of inoculum due to continuous planting, sufficient populations of active GLH vectors, and favourable environmental conditions contributed to the tungro epidemics. All of these factors, including the continuous availability of rice at a susceptible growth stage in intensively cropped areas, must be present concurrently for epidemics to occur. In recent years, although no major outbreak has been reported, the disease has become endemic in some areas, resulting in sporadic damage. These areas are often infected with tungro and inoculum sources abound, creating a high tungro inoculum potential which could lead to major epidemics.

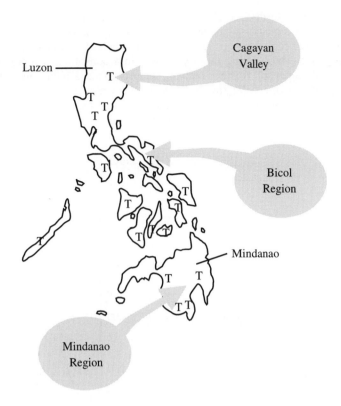

Figure 1   Tungro occurrence (T) in the Philippines in 1993–94 showing three 'hot spot' areas.

Reported outbreaks

'Tungro', which means degenerated growth in the Philippine language, was first observed in the Philippines at the experimental farm of the International Rice Research Institute (IRRI), Los Baños, in 1963. RTD was first shown to be a transmissible disease by Rivera and Ou (1965), although the disease is thought to have occurred in the 1940s in the Philippines and even earlier elsewhere. RTD is very important and dangerous because of its potential to cause severe epidemics and total yield loss over large areas, and also because it is unpredictable. Major outbreaks of RTD occurred in 1957, 1962, 1969, 1971, 1975, 1977 and 1983–84. According to Serrano (1957), the *'accep na pula'* or stunt disease, which is now considered to have been tungro, was extremely destructive in the 1940s throughout the major rice-growing regions of the Philippines and caused 30% loss overall, which was equivalent to 1.4 million t, each year.

In 1971, when the early IRRI varieties IR5, IR8, IR22 and IR24 were widely cultivated, yield losses due to RTD in the Philippines were estimated to be as high as 456 000 t of rough rice. The disease has also caused severe damage in many parts of the region since 1980. Major rice-growing areas including the Central Plain of Luzon and parts of Visayas and Mindanao Islands were severely affected by tungro when IR8, IR36 and IR42 were planted widely in the Philippines in 1984. High tungro incidence was also observed in South Cotabato and Cotabato Provinces in 1985 and 1986.

Occurrence in some hot spot areas

Bicol Region (1983–89): The largest rice growing areas are in Camarines Sur and Albay Provinces and — particularly in Polangui, Albay — are characterized by continuous planting with no strict fallow period and as many as three rice crops per year. Moreover, planting dates differ widely from farm to farm. Almost all rice growth stages are present throughout the area during the season. Large areas were affected by RTD between 1983 and 1989, with a peak of 1981 ha in the 1989 dry season (Figure 2a). Because planting is continuous, RTD occurred throughout the year. In both 1988 and 1989, peak tungro incidence occurred in April (Figure 2b). In all areas affected by tungro, the major varieties used by the farmers were IR36, IR42, IR64, IR60, IR72 and BPI Ri10.

Mindanao Region (1988–93): North Cotabato on the island of Mindanao is considered the most compact province in terms of land mass, including an extremely large, flat, contiguous lowland rice area. Previously, two to three crops per year were grown routinely, but this situation has changed due to lack of sufficient irrigation water during the dry season. In a consolidated report of the Department of Agriculture based in Kidapawan, North Cotabato, tungro-affected areas amounted to more than 2200 ha in 13 municipalities from 1989 to 1993. Davao Province is located in the central part of Mindanao and grows much of the rice produced in the region. Department of Agriculture reports revealed that the total area affected by tungro in the province was 2100 ha between 1988 and 1993.

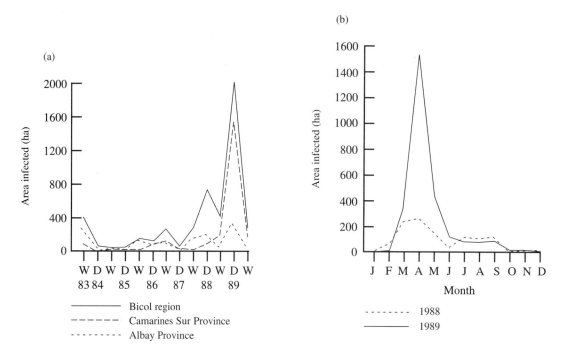

Figure 2   Tungro damage in Bicol Region (a) in wet and dry seasons, 1983–89 and (b) monthly in 1988 and 1989.

Partially reported outbreaks in 1993

Davao del Norte: Partially consolidated reports for 1993 on RTD incidence revealed that a total of 50 328 cavans (1 cavan = *c*. 45 kg rough rice) amounting to 10 588.853 pesos (Ps 27 = US$ 1.00) were lost due to the disease. These rough estimates suggest that 71% of the rice area in the province was affected which is the highest recorded incidence of tungro in the area. All municipalities were affected, with high incidences in Sto Tomas (271 ha), Nabunturan (158 ha), New Corella (127 ha) and Montevista (21 ha).

Davao del Sur (Digos, Matanao and Santa Cruz): A high incidence of RTD was recorded in September 1993 in Matanao, Davao del Sur Province. Seven *barangays* (smallest local government unit) were heavily infected with tungro, and 517 ha were damaged. RTD incidence ranged from 34% to 90% and yields from 5 cavans/ha to 38 cavans/ha. Farmers were advised to remove (rogue) infected plants and plough-in heavily infected fields to reduce the amount of inoculum present. The major varieties planted included IR64, IR60, IR72, and Masipag selections and PSBRc series, which are newly-approved and released varieties by the Philippine Seedboard (PSB).

CURRENT TUNGRO RESEARCH AND DEVELOPMENT ACTIVITIES

Since the mid-1960s much of the research and development work on RTD in the Philippines has been done by IRRI. No other institution in the country did any intensive research on the disease until PhilRice became fully operational in 1992. The current tungro research activities of PhilRice on RTD are outlined.

Varietal improvement

The objectives of the varietal improvement programme are: (a) the development of stable, high-yielding varieties for specific agro-ecological zones, especially for 'hot spot' areas; (b) the improvement of current PSB varieties by incorporating appropriate tungro resistance genes through conventional and non-conventional methods; (c) the development of direct-seeded rice varieties highly resistant to RTD. These objectives are being realized through the following projects.

- **Breeding programme at PhilRice** Lines or selections developed at PhilRice are field-tested for their reaction to pests and diseases under field conditions. The material passes through the observational nurseries (ONs) and the preliminary yield trials (PYTs). The advanced lines derived in this way, which usually have high yield potential, acceptable pest and disease ratings and good grain quality, are passed on to the National Rice Co-operative Testing Project (NRCTP) under the Rice Varietal Improvement Group.

- **National Co-operative Testing Project** This project covers breeding lines developed by several institutions including IRRI, PhilRice and the University of the Philippines in Los Baños (UPLB). PhilRice co-ordinates the project and consolidates data from different stations throughout the country. The promising lines are tested in different locations for yield and agronomic acceptability, pest and disease reactions, and grain quality. After several seasons of testing, all promising lines that show very good qualities are recommended for release as PSB varieties.

- **Physiological specialization of rice tungro viruses** This is a joint collaborative effort between PhilRice and IRRI. Thirty accessions from the International Rice Germplasm Collection (IRGC) at IRRI found to be resistant to RTSV in tests at IRRI are being screened in two locations to assess possible differential reactions to the strains of RTSV present in the respective areas (see Koganezawa and Cabunagan, page 54). This project started in the 1993 wet season and continued until late 1994.

- **Field-testing of mutant lines or selections for tungro resistance** This is another IRRI–PhilRice collaborative project, which is being conducted at the PhilRice Midsayap Experiment Station (MES) in Mindanao. Lines or selections, which had been subjected to mutagens, are being field-tested for their reaction to tungro viruses occurring in the area.

Cultural management and crop production systems

Crop production systems associated with tungro management and level of farmer adoption

An operational management scheme for RTD with several component technologies is being tested in some 'hot spot' areas. The component technologies include the combination of time of planting and use of vector-resistant varieties, and direct-seeding is being compared with transplanting. The use and proper timing of insecticide application is also being assessed to avoid tungro and its secondary spread.

Screening of synthetic pyrethroid insecticides and their effect on GLH and RTD

In an experiment conducted in 1989–90 at the PhilRice MES, pyrethroid-treated plots had significantly lower GLH populations and less tungro infection than untreated controls (Batay-an, 1992). Cypermethrin and ethofenprox had decreased GLH populations and tungro incidence at all sampling dates. Higher yields were obtained in treated compared with untreated plots. Moreover, foliar sprays of cypermethrin were safe to spiders and other natural enemies such as *Cyrtorhinus lividipennis* Reuter and *Agriocnemis pygmaea* (Rambur) when applied 20 days after transplanting (DAT). None of the pyrethroids affected natural enemy populations at 35 DAT. Extracts of Makabuhai (*Tenespora crispa*), a local plant, were tested against GLH but were ineffective in controlling populations. Yields from Makabuhai-treated plots were comparable with pyrethroid-treated plots, except cypermethrin where the yield was higher.

Socio-economics of RTD and farmers' perceptions and attitudes towards the disease and its control

The following research activities are being conducted under this programme:

- survey, document and characterize of tungro epidemics in 'hot spot' areas
- determine demographic profile of farmers in tungro 'hot spot' areas
- assess farmers' perceptions and the control measures currently adopted.

A survey on farmers' perceptions and current tungro control measures was made in Bukidnon Province, Mindanao, in August 1993. A total of 79 farmers in rainfed and lowland rice areas were interviewed. The findings were considered under three headings.

- **Control practices of farmers if tungro is present in their field** The results showed the reliance of farmers on chemicals once RTD occurs. The majority of farmers (87%) said they would use chemicals (spray or granular application) at least once to control RTD. Of these, a high proportion stated they would spray insecticides. Surprisingly, and erroneously some farmers in rainfed areas would resort to the application of the fungicide Hinosan (13% of those interviewed) or a herbicide (6%) to control tungro. Moreover, 24% said they may remove (rogue) diseased hills from their fields. Of these, 10% would then burn the rogued plants to reduce the amount of inoculum available. Other control options referred to by respondents included water control (2%) and the use of dapog seedlings (1%). Dapog seedlings are those raised in seed-beds either in the field or in a concreted area at a very high rate (100 kg/ha) and they are ready for transplanting within 9–14 days.

- **What farmers would do if tungro appeared the following season** Many farmers (49%) said they would apply chemicals. Of these, 47% would resort to 'massive' (community-based) spraying using insecticides but no specific chemicals were mentioned. Farmers would rogue (28%), change variety (5%), inform a technician (3%), destroy diseased plants (1%), take no action (1%), apply granules (1%), and apply fertilizer (1%).

- **What farmers would do if the neighbouring farmer has tungro** Most of the respondents (78%) said they would apply chemicals, and 8% of these would use a fungicide (Hinosan) and 5% a herbicide (Machete or Rogue). A few respondents (1%) said they would apply fertilizer. The remaining farmers stated they would advise their neighbours to apply insecticide. Some farmers (19%) said they would destroy diseased plants while others (3%) opted to do nothing.

The survey showed that farmers' knowledge of RTD is inadequate. Control options are largely limited to chemical treatment because other measures are inappropriate or are impractical and often require too much labour. Tungro is still considered as potentially damaging and farmers are risk-averse when dealing with this disease. There is a need for further training and information on all aspects of the disease. Moreover, improved tungro management options are required for recommendation to farmers.

INFORMATION SYSTEMS AND TECHNOLOGY TRANSFER

Rice technology development and utilization

A programme is being initiated involving the intensification and expansion of location-specific on-farm technology demonstration efforts for tungro management. This will be closely supported by technology promotion through training and communications programmes and is based on the assumption that appropriate technologies for tungro management are available for a given location. In locations where such technologies are available, they should be technically and socially acceptable. Thus, the rice R&D system should aim to upgrade the technical knowledge, skills and attitude of less efficient farmers to approach the efficiency levels of the better farmers regarding management of tungro and overall production.

Training and technology transfer

Training is being developed to support the Agricultural Training Institute (ATI) — the training arm of the Department of Agriculture — and also to meet the requests of cooperatives and other non-governmental organizations (NGOs). Tungro technology demonstration efforts will be promoted among NGOs and local government units (LGUs). Likewise, training will be provided for those who will train field technicians and farmers, and for other specialists who may be needed in the implementation, monitoring and evaluation of the tungro research and training components. Training will be by (a) module type, comprising tungro management guidelines, trainers' training, technical briefing and specialized programmes, and (b) client type, involving trainers, farmers and farmer-leaders, agricultural technologists, researchers, NGOs, broadcasters and programme managers.

Farmer empowerment by enhancing the decision-making of farmers and improving communication tools will be done by training (a) technicians of the sustainable rice production programme with NGOs, involving an integrated tungro management scheme, (b) integrated tungro management specialists and trainers to support village or community-level tungro management projects in 'hot spot' areas in the Philippines, and (c) extension workers who will be involved in research and field demonstration of promising tungro technologies in applied research and farming systems.

Communication support tools for technicians and network members

The following activities are being conducted: (a) development, production and pre-testing of video and print materials for extension workers and farmers; (b) preparation of an illustrated technical guide in various appropriate languages; (c) production of rice technology leaflets and press releases and radio broadcasts.

RECOMMENDATIONS FOR TUNGRO CONTROL AND MANAGEMENT OPTIONS FOR FARMERS

The following control options are listed as necessary to manage tungro in perceived 'hot spot' areas in Mindanao, based on a report from the provincial office of the Department of Agriculture:

- **Create a crop-free period of at least six weeks between crops** This practice is intended to break the cycle of the GLH vectors in the rice habitat and to reduce the inoculum potential for the next cropping season by destroying all previously infected rice stubbles.

- **Avoid excessive application of nitrogenous fertilizer** The widespread practice of applying nitrogen as a response to tungro has no direct effect on the spread or severity of the disease, especially when late infection occurs. Some farmers may claim that, in early tungro infections, a minimal greening effect on plants may appear but may not increase yield even after nitrogen application. Leaf number may increase significantly but leaves may soon die. So far, no scientific evidence is available to explain this although research continues.

- **Use resistant varieties such as IR60, IR62, IR70 and IR72** This recommendation may be location-specific. The varieties may have been largely unaffected by tungro in some areas, but not necessarily in others. Thus, it is important to determine the varieties most suitable for endemic areas and to make appropriate recommendations. This measure alone may not suffice to ensure sustained control.

- **Plant synchronously** This measure has been recommended for many years but it is usually difficult to implement for socio-cultural reasons. Close co-ordination between government agencies such as the Department of Agriculture, National Irrigation Administration, local government units and NGOs is needed to implement and sustain this recommendation.

- **Create farmers' awareness of the recommended economic threshold level (ETL) of 1–2 GLH/ 10 hills for spraying insecticides. Alternatively, incorporate systemic granular insecticides into the soil before sowing the seed-bed** Chemical applications may be unnecessary in areas planted synchronously and as long as the peak of the GLH population does not occur in the first 40 DAT. Where preventive chemical control is recommended, basal application of a systemic granular insecticide, carbofuran (Furadan 3G), should be done in the seed-bed before sowing the seeds. Mipcin should be applied if GLH numbers exceed the economic threshold applied.

- **Discourage farmers from ratoon practices** After harvesting, farmers do not immediately destroy crop stubbles. Ratoons of previously-infected rice crops appear in the field when soil moisture is adequate and serve as inoculum sources for the next crop. The infected crops should be destroyed by ploughing them under the soil. When tungro occurs in the new crop, all infections should be rogued continously and the removed plants burned to destroy the inoculum.

RESEARCH AND TECHNOLOGY CONCERNS FOR MANAGEMENT OF RICE TUNGRO DISEASE

Sanitation

Sanitation and in particular roguing is used by some farmers to restrict tungro to a low level. Some farmers burn the infected hills after roguing; however, they generally find roguing time-consuming and tedious. It may also require additional labour.

Heavily infected fields should ideally be ploughed under. However, for economic reasons, subsistence farmers seldom do this and simply abandon the field which then becomes a source of inoculum from which spread occurs to contemporaneous or later crops.

Insecticides

Insecticides have been widely used by Filipino farmers, both as eradicants and protectants, but farmers are more afraid of tungro disease than the GLH vectors. Indeed, from some interviews conducted, it became apparent that farmers readily spray when tungro symptoms appear in their field with the object of curing the yellowing and stunting symptoms, rather than controlling GLH.

Routine spraying is done as part of the normal cropping system. Emergency spraying is done after the appearance of tungro symptoms and is targeted against the vector.

Seed-bed application of insecticide is in the form of granules, specifically carbofuran (Furadan 3G), which is claimed by farmers to provide protection during the first 30 days since the chemical is absorbed by the plant.

Resistant varieties

Commercial rice varieties that are available are mostly vector-resistant. Varieties resistant to both tungro viruses have not been developed, although they have been sought. Previous accounts stress the importance of introducing resistance to both viruses because resistance to the GLH vectors can be easily incorporated but easily breaks down.

Recent findings have confirmed the existence of a strain of RTSV in Mindanao which can infect some varieties previously identified as resistant. This finding will help focus future research on the need to breed for durable resistance.

OTHER RESEARCHABLE AREAS

- Cropping practices: seeding method, planting time/date, synchronization of planting, field isolation/ disposition
- characterization of 'hot spot' areas in the Philippines — establish criteria for classification and evaluation; determine factors contributing to tungro endemicity in certain areas
- varietal rotation scheme and its components
- alternative control options focused on the vectors; field ecology of GLH and population dynamics
- develop innovative screening techniques to evaluate durable/stable types of resistance
- pyramiding major genes
- genetic variability of the vectors and viruses
- rapid methods of disease detection in farmers' fields
- database or tungro profile in the Philippines
- socio-economic aspect of tungro and its management in the Philippines
- epidemiological aspects or spatial/temporal spread of the disease
- biological control of insect vectors.

REFERENCES

BATAY-AN, E.H. (1992) *Effect of Synthetic Pyrethroid Insecticides on GLH and its Natural Enemies and Tungro*. Terminal Report. Mindanao: Philippine Rice Research Institute, Midsayap Experiment Station.

RIVERA, C.T. and OU, S.H. (1965) Leafhopper transmission of 'tungro' disease of rice. *Plant Disease Reporter*, **49**: 127–131.

SERRANO, F.B. (1957) Rice *accep na pula* or stunt disease — a serious menace to the Philippine rice industry. *Philippine Journal of Crop Science*, **86**: 203–23.

Other Publications

CHOWDHURY, A.K., TENG, P.S., HIBINO, H and MEW, T.W. (1989) Some aspects of rice tungro epidemiology. *IRRI Saturday Seminar, 21 October 1989.*

CIPRIANO, T.S. Jr. (1993) *Field Report on Tungro Occurrence in Polangui, Albay.* Polangui, Albay, Philippines: Municipal Agricultural Services Office.

DEPARTMENT OF AGRICULTURE (Region XII). *Report on the Status of RTV infestation in Davao del Sur and Davao del Norte*, Davao City, Philippines: Department of Agriculture.

INTERNATIONAL RICE RESEARCH INSTITUTE (1971) *Annual Report for 1970.* Manila, Philippines: International Rice Research Institute.

LING, K.C. and PALOMAR, N.K. (1966) Studies of rice plant infected with tungro virus at different ages. *Philippine Agriculture,* **50**: 165–177.

OFFICE OF THE PROVINCIAL AGRICULTURIST (1993a). *Rice Tungro Infection Report in North Cotabato.* Amas, Kidapawan, Cotabato, Philippines.

OFFICE OF THE PROVINCIAL AGRICULTURIST (1993b) *Rice Tungro Infection Report for Davao del Sur.* Digos, Davao del Sur, Philippines.

REGIONAL CROP PROTECTION CENTER (Region 5). (1993) *Tungro Status Report.* BRIARCS, Pili, Camarines Sur, Philippines.

PHILIPPINE RICE RESEARCH INSTITUTE (1989) University of the Philippines at Los Baños, College, Laguna, Philippines.

PHILRICE (1993). *Training for Farmer Empowerment: The PhilRice Experience.* Maligaya, Munoz, Nueva Ecija, Philippines.

Paper 11

Rice tungro disease and green leafhopper vectors in Nepal: current status and future research strategies

G. DAHAL*

Institute of Agriculture and Animal Sciences, Tribhuvan University, PO Box 984, Kathmandu, Nepal

INTRODUCTION

Nepal is an agricultural country and the agriculture sector contributes *c.* 56% of gross domestic product. Because of the mountainous terrain only *c.* 18% of the total land area is available for cultivation, but diverse farming is practised in different regions. Farming is mainly of the subsistence type and yields per unit area of land are very low. Rice is the most important food crop of Nepal and is grown on 60% of the total cultivated land in all agro-ecological zones ranging from the Terai (100–300 m), valleys and foothills (100–1000 m), to the high mountains (2600 m) (Sthapit, 1992). Terai is the principal agricultural region representing *c.* 70% of the total cultivated land area. Double cropping of rice ceases at *c.* 900 m and rice is not grown above 2600 m. Few countries have such a diversity of rice growing environments as Nepal where the crop is grown in five major eco-systems (Table 1).

Table 1 Major rice farming systems in Nepal

Type	Local name	Environment area	Planting/harvesting details	% of total rice
Early rice	*Chaite/Judi/ Heude dhan*	Low altitude (100–700 m); assured irrigation	Seeded in February, transplanted in the Nepali month of *Chaite* (Mar/Apr) and harvested before mid-July	10
Main season rice		100–2600 m altitude; partial or full irrigation	Sown in May/June and later transplanted after *Chaite* rice is harvested (mid-July onwards)	52
High altitude rice		1000–2600 m altitude; rainfed or partial irrigation	Sown from March to May depending upon altitude and usually transplanted in June–July. Partially irrigated and rainfed terraced systems are used	26
Upland rice	*Ghaiya*	Totally rainfed in low altitude (300–800 m)	Directly sown in unbunded fields of the plains area as soon as rain occurs. Seeding usually from March–April and harvesting in late August	9
Deep water rice		Ghol area of Terai (<200 m common in low lying areas of river basins and plain areas of Terai		3

Source: Sthapit, 1992.

Rice contributes 52% of the total grain production of the country and dominates the agricultural sector, engaging over 75% of the working population (Garrity *et al.*, 1987). Currently, from *c.* 1.4 million ha of land, nearly 3.5 million t of rice are produced. Although rice yields are high in some parts of the country, they are very low in many places and national average yields have remained static at 2.0 ± 0.48 t/ha for the past 20 years (International Rice Research Institute, 1990). Various constraints have been identified and rice diseases are considered important.

---

*Present address: International Institute for Tropical Agriculture (IITA), Oyo Road, PMB 5320, Ibadan, Nigeria.

More than 50 diseases affect rice in Nepal and based on their incidence, distribution and effect on yield or yield components about eight are considered economically important (Dahal *et al.*, 1992). So far, two virus diseases, rice tungro and rice dwarf, have been identified (John *et al.*, 1979; Omura *et al.*, 1981). Rice tungro (Ling, 1972) which was first observed in the Philippines (Rivera and Ou, 1965) is now recorded as the most important virus disease of rice in South and South-East Asia. Rice tungro virus disease (RTVD) or rice tungro disease (RTD) is popularly known as 'tungro' or '*rate rog*' (red disease) in Nepal. Although the disease has caused epidemics and considerable economic loss in many South and South-East Asian countries including Bangladesh, Philippines, Thailand, Malaysia, Indonesia, and India (Ou, 1985; Ling and Tiongco, 1980), it is not considered important in Nepal because of its low incidence or irregular occurrence in isolated areas. Nevertheless, because of the prevalence of the vector leafhoppers *Nephotettix virescens*, *N. nigropictus* and *Recilia dorsalis* (John *et al.*, 1979; Sharma and Mathema, 1976), together with the cultivation of susceptible rice varieties in large areas and the occurrence of RTD in nearby areas of West Bengal and Bihar in India, the disease is considered a threat to rice production in Nepal. This paper reviews past research on RTD and its leafhopper vectors in Nepal, highlights ongoing research and suggests strategies for future research.

RICE TUNGRO DISEASE

Occurrence

Rice tungro disease was first reported in Nepal from Parwanipur, Bara district, where characteristic symptoms occurred in early to mid-tillering stages of rice crops (John *et al.*, 1979). RTD-like symptoms were then observed in rice cultivar IR20 in Dupery Panchayat at Janakpur, Dhanusha district (Mallik, 1981). Later, Omura *et al.* (1981) collected rice plants showing yellow discoloration and stunted growth from Hardinath Farm, Janakpur. Based on the results of electron microscopy, they confirmed the presence of both rice tungro spherical virus (RTSV) and rice tungro bacilliform virus (RTBV) in the infected leaf tissues. However, there are no reports of RTD occurring in large areas or causing severe epidemics in Nepal. Recently, Dahal *et al.* (1993) surveyed most rice growing areas of the southern Terai plains of Nepal and confirmed the presence of both RTBV and RTSV in the Parwanipur and Hardinath areas (Table 2).

Symptomatology

Symptoms of RTD-infected rice in Nepal are yellow leaf discoloration, stunted growth, delayed flowering, and slightly reduced tillering (Omura *et al.*, 1981; Dahal *et al.*, 1993). Detailed symptomatological observations of rice plants infected by RTBV and/or RTSV are currently in progress. In Indonesia, plants of RTD-susceptible rice cultivars infected with both RTBV and RTSV were severely stunted and discoloured, whereas the plants infected with RTBV alone were moderately stunted, usually with dark green leaves which were occasionally yellow orange (Hibino *et al.*, 1978). Plants infected with RTSV alone developed no clear symptoms and were not stunted or discoloured.

Table 2 Incidence of rice tungro-like diseases during the main rice season in two districts of Nepal in 1993*

District/Location	Cultivar	Disease incidence (%)	Leafhopper (no./30 sweeps)	Major symptoms†
Janakpur: Hardinath	Sabitri	10	12.2	Rt St y
	Masuli	–	3.6	–
	Makawanpur-1	28–30	33.3	Rt St y
Bara: Parwanipur	Sabitri	–	25.0	–
	Masuli	70–80	100.0	Rt St Y

Source: Dahal *et al.*, 1993.

*Survey conducted at crop booting stage in research plots covering *c.* 50 ha of rice. From each diseased patch suggestive of tungro, *c.* 100 hills were examined and incidence was determined; depending on the incidence of the disease, 1–5 such observations were taken from each plot or variety.

†Rt = reduced tillering, St = mild stunted growth, y = yellowing of plants with very mild orange yellow leaf discoloration, Y = yellowing of plants with very distinct orange yellow leaf discoloration.

Diagnosis

As in most other developing countries where laboratory facilities are limited, correct diagnosis of RTD-infected plants has been a problem in Nepal where the symptoms have often been confused with those due to soil mineral deficiencies. However, this problem can be overcome by using diagnostic techniques such as the latex agglutination test (Omura *et al.*, 1984), or the modified starch test (International Rice Research Institute, 1983) which can be used effectively, even where laboratory facilities are not available. Recently, enzyme-linked immunosorbent assay (ELISA) (Bajet *et al.*, 1985) has been introduced to index plants for rice viruses. Facilities for serological detection were established recently at the Institute of Agriculture and Animal Sciences (IAAS), Rampur, Chitwan, Nepal.

Aetiology

Electron microscopy of Nepalese virus isolates from tungro-diseased plants was conducted by Omura *et al.* (1981) in Japan who reported that rice plants with yellow leaf discolouration and stunted growth collected from Hardinath farm, Janakpur, contained polyhedral particles of *c.* 30 nm in diameter (RTSV) and bacilliform (RTBV) particles 30–35 nm wide and 100–300 nm long. From this study, it was confirmed that RTD in Nepal was caused by both RTBV and RTSV, as reported in the Philippines, Thailand, Malaysia and Indonesia. Recently, Dahal *et al.* (1993) indexed RTD isolates collected from Hardinath and Parwanipur (Bara District) by latex agglutination test (Omura *et al.*, 1984) and ELISA (Bajet *et al.*, 1985) and showed that both isolates reacted with RTBV and RTSV antiserum. Further tests using the polymerase chain reaction (PCR) confirmed the presence of RTBV. Features of the PCR product of RTBV closely resembled those of various Indian rice tungro isolates and differed from those from the Philippines (Dahal *et al.*, 1993). In preliminary cross-hybridization studies, the RTBV DNA of a Nepal isolate hybridized more strongly with an Indian tungro isolate than with the Philippines tungro isolate (Dahal *et al.*, 1993).

Disease cycle

In Nepal, only three leafhopper vector species (*N. virescens, N. nigropictus* and *R. dorsalis*) are reported on rice and it is assumed the transmission of tungro viruses is due to these species. Because of the semi-persistent nature of tungro viruses in their vectors, the insects have to acquire these viruses from a sequence of diseased plants to remain infective for long. Thus the infectivity of vector populations is influenced by the prevalence and potency of virus reservoirs present in the area. Hence, RTD is most serious in tropical Asian countries where the host plants grow and insect vectors reproduce throughout the year (Ou, 1985). In areas, such as Nepal where rice is generally grown only once a year during the summer season (June–July to October–November) and vector populations decrease during the winter season (November to mid-March), the viruses can survive in rice stubble, or in alternative hosts, especially weeds which are probably important sources of inoculum for subsequent seasons.

Many weed and other plant species have been reported as hosts of tungro viruses (Wathanakul, 1964; Rivera *et al.*, 1969; Rao and Anjaneyulu, 1978; Anjaneyulu *et al.*, 1982) but there are no reports on the host range of tungro viruses from Nepal. Neither are there reports of studies related to screening Nepalese indigenous rice germplasm against rice tungro viruses.

Ongoing research

Studies on identification and characterization of rice virus diseases in Nepal based on biological, serological and molecular characteristics are currently in progress at IAAS, Rampur, and John Innes Institute (JII), Norwich, UK, under a project '*Studies on virus diseases of rice and their vectors in Nepal*' supported by the Rockefeller Foundation. The project seeks to (a) determine the current status of rice virus diseases in Nepal, (b) isolate and characterize representative virus isolates, and (c) determine the virus/leafhopper/rice interactions. Current progress includes isolation and preliminary transmission of tungro viruses, cloning of the RTBV genome and comparative molecular analysis with other South and South-East Asian tungro isolates. The studies on rice tungro viruses and their leafhopper vectors were intended to link with the recently completed UK Overseas Development Administration/Natural Resources Institute-funded project on '*Variability in rice tungro viruses*' at JII, Norwich, UK.

Table 3   Reports of rice green leafhoppers in rice fields during 1972–90 in Nepal

Year	Cultivars	Monitoring methods*	Locations	References
1972	NA	–	Kathmandu	Pradhan and Joshi, 1972
	NA	SW	Kathmandu	Joshi, 1972
	IR8 IR20	SW	Palpa	Joshi, 1972
1974	NA	SW	Palpa	Sharma and Khatri, 1977
1976	NA	SW	Kathmandu, Parwanipur	Adhikari *et al.*, 1986
1976–77	TN242	SW	Khumaltar	Sharma and Khatri, 1977
	NA	LT	Khumaltar	Pradhan, 1980
1979	NA	SW	Khumaltar	Pradhan and Khatri, 1980
1980	NA	?	Khumaltar	Pradhan and Khatri, 1980
1982	NA	LT	Nepalganj	Palikhe, 1982
	NA	SW	Nepalganj, Dhankutta, Chitwan	Pradhan, 1983
1989	Masuli	SW/LT	Parwanipur, Rampur	Jyoti, 1989
1990	Masuli	SW	Parwanipur, Chitwan, Hardinath	Jyoti, 1990

*SW = sweep-net catch, LT = light-trap catches.

RICE GREEN LEAFHOPPERS

Occurrence and damage caused

Regular surveys of insect pests of rice, including rice green leafhoppers (GLH), have been carried out by the Central Entomology Division, Nepal Agriculture Research Council, Khumaltar since 1969 (Pradhan, 1989) and the major records of leafhopper occurrence are summarized in Table 3. Sharma and Mathema (1976) identified 11 species of leafhoppers and planthoppers including *N. nigropictus* (syn. *N. apicalis*) and *R. dorsalis* (Syn. *Inazuma dorsalis*) from limited samples collected from Kathmandu valley and Biratnagar. These leafhoppers have also been reported from Palpa, Chitwan, Kathmandu, Biratnagar, Nepalganj, Jhapa and Dhankutta districts.

Biological studies

The results of previous studies on the biology of *N. nigropictus* (Pradhan, 1973; Pradhan and Joshi, 1972; Sharma and Mathema, 1976) are summarized in Table 4. Under laboratory conditions, the eggs took an average of 11.6 days to hatch and the nymphal period was *c.* 23.8 days (Pradhan and Joshi, 1972). Further studies conducted under laboratory and field conditions indicated a nymphal period of 24.5 days (Pradhan, 1973). The large variation in nymphal period observed in these studies could be due to climatic differences as these studies appear to have been conducted under the variable temperature regimes encountered under natural conditions. Recent results (Dahal *et al.*, unpublished data) on the biology of *N. nigropictus* and *N. virescens* on rice cultivars Masuli and TN1 indicate that the nymphal period of both species was *c.* two weeks, but the life span of adult *N. virescens* was slightly longer than that of *N. nigropictus* (Table 4).

Table 4   Biology of rice green leafhoppers in Nepal

Leafhopper species Cultivar	Duration of different stages in days								
	Egg	1st	2nd	3rd	4th	5th	Total nymphal period	Adult lifespan	Source*
N. nigropictus									
TN242	–	3.0	2.7	2.5	5.6	2.0	15.8	16.0	1
TN1	11.6	–	–	–	–	–	23.8	22.5	2
NA	–	4.9	3.8	4.4	6.3	5.4	25.4	–	3
TN1	–	2.5	2.6	3.2	3.3	3.5	15.1	11.6	4
Masuli	–	1.7	2.8	2.8	2.5	3.3	13.2	11.5	4
N. virescens									
TN1	4–11	2.0	3.3	3.2	3.3	3.2	15.0	19.5	4
Masuli	4–11	2.0	3.2	2.7	2.5	3.7	14.1	16.2	4

*1 = Sharma and Mathema, 1976; 2 = Pradhan and Joshi, 1972; 3 = Pradhan, 1973; 4 = Dahal *et al.* (unpublished data).

Population dynamics

Population dynamics were studied of rice green leafhoppers, caught in light traps at Kathmandu valley (Khumaltar, 1360 m) and at three relatively low-altitude sites (200–250 m): Janakpur (Hardinath), Bara (Parwanipur) and Chitwan (Rampur). At Khumaltar, there was a steady rise in GLH populations from April to a peak in October, except in 1976 when the peak population occurred during September. The population then decreased gradually to a minimum during December (Pradhan, 1979). At Hardinath, the leafhopper populations had two peaks: during July–August and October–November. The populations then decreased to a minimum during December and January (Aota, 1979). At Parwanipur, GLH population was highest during 1984 and low during 1981. Each year maximum populations occurred during September–October (Adhikari *et al.*, 1986). In 1985, the leafhopper population had two peaks: a small peak during July and the main peak in late October. At Rampur, the leafhoppers had two peaks: a small peak during mid-June to July and the maximum peak during mid-October. The leafhopper population declined sharply after the end of November and few leafhoppers were caught during the winter months (December and January) when minimum temperatures were 2–5°C.

A comparison of leafhopper catches between 1989 and 1992 indicated that the general population trends were similar each year but with fluctuating peak periods. During 1989, the leafhoppers had a major peak during mid-November, while in 1992 the leafhoppers had two distinct peaks; a small one occurred during mid-July and a second in mid-October — about one month earlier than in 1989. Results of leafhopper monitoring in these four locations indicated that population build-up at Khumaltar occurred earlier than at the three sites at lower altitudes on the southern Terai plains, where the main rice season is generally four to six weeks later than at Khumaltar.

Ongoing research

Studies are in progress on population dynamics, species composition and characterization of leafhopper populations from different ecological zones by biological, morphological and biochemical characteristics. Preliminary information on some of these topics is summarized below.

Population density and species composition

Previous studies on population densities of leafhoppers were not focused on determining populations of *Nephotettix* species. From limited field sampling data, John *et al.* (1979) and Inoue (1986) reported *N. nigropictus* as the dominant leafhopper species in Nepal. In recent studies, Dahal *et al.* (unpublished data) determined the species composition and population densities of *Nephotettix* spp. from rice fields during September–October at maximum tillering stage from 70 locations covering 20 districts of Nepal, mostly located in the southern Terai plains. In most of the locations, both *N. virescens* and *N. nigropictus* were collected but their composition varied considerably — *N. virescens* was usually the more numerous. The ratio of *N. virescens* to *N. nigropictus* varied from 0.2 to 39.8; the highest ratio was recorded in Siraha (Table 5). Irrespective of the collection sites, the overall mean ratio of *N. virescens* to *N. nigropictus* was variable depending on cultivar and altitude.

The populations of *N. virescens* and *N. nigropictus* varied depending on the rice cultivars grown and the location (Table 5). At all sites, except those in the mid-hills of Kaski, Makawanpur, Syanja and Kabre, the populations of *N. virescens* were much higher than those of *N. nigropictus*. The population of *N. virescens* was highest in Morang and Siraha, while the populations in other locations were generally low. In the valleys where mostly improved cultivars were planted, densities of *N. virescens* were higher than those of *N. nigropictus*. By contrast, in the mid-hills where mostly local cultivars were planted the population density of *N. nigropictus* was higher.

In a separate study, leafhopper populations were monitored on nine improved cultivars planted at the IAAS experimental farm. The populations of *N. virescens* and *N. nigropictus* and their relative numbers varied depending on cultivar (data not shown in table). The ratio of *N. virescens* to *N. nigropictus* was highest on Makawanpur-1 and Radha-6, and lowest (2.4) on Bindeswari. On cultivars such as Sarju-52 and Sabitri, the populations of *N. nigropictus* were comparable with those of *N. virescens*.

Table 5 Number of leafhoppers collected by sweep nets from rice fields planted with improved or local cultivars from 20 districts of Nepal (Dahal *et al.*, unpublished data)

Districts: altitude	Rice cultivar	No. of fields	Leafhoppers (no./30 strokes)				Ratio*
			N. virescens		*N. nigropictus*		
			Range	Average	Range	Average	
Improved cultivars							
Bara: 200 m	Bindeswari	1	0–4	1.43	0–3	0.29	4.90
	Chaite–2	1	1–6	3.43	0–2	0.57	6.02
	Chaite–4	3	0–3	5.29	0–2	0.37	14.30
	Ghaiya–2	1	0–4	1.13	0–1	0.25	4.52
	HIT–44	1	0–1	0.10	–†	0.00	–
	Janaki	1	0–2	0.46	0–1	0.27	1.70
	Makawanpur	1	4–7	5.60	1–3	1.60	3.50
	Masuli	3	0–7	2.35	0–2	0.98	5.90
	Radha–7	1	0–3	0.83	0–1	0.25	3.32
	Radha–9	1	0–2	0.40	0–1	0.10	4.00
	Radha–17	1	0–3	1.60	–	0.00	–
	Radha–32	1	0–3	1.63	0–2	0.75	2.17
	Sabitri	1	2–4	1.20	0–1	0.40	3.00
Bhaktapur: 1380 m	Masuli	1	–	0.00	0–3	1.60	–
Chitwan: 228 m	15015	1	2–7	5.00	0–3	1.40	3.57
	CH–45	1	0–1	0.33	0–1	0.30	1.10
	Masuli	11	0–41	17.96	0–12	4.30	4.18
	Pant–4	1	3–6	4.40	0–2	1.00	4.00
	Radha–9	1	7–12	9.00	1–4	2.40	3.75
	Radha–17	1	7–10	8.60	2–5	3.60	2.39
Dhading: 400–1040 m	Himali	1	0–8	4.60	1–4	2.60	1.80
	Masuli	4	4–16	9.80	1–9	4.80	2.04
Janakpur: 200 m	Makawanpur	1	0–1	0.30	0–1	0.10	0.30
	Masuli	1	0–1	0.30	0–2	0.40	0.75
	Sabitri	1	0–3	0.40	0–1	0.10	4.00
Jhapa: 200 m	Masuli	4	18–125	64.70	2–8	4.10	15.78
Kabhre: 1450 m	Khumal–4	1	0–1	1.00	1–2	1.00	1.00
	Masuli	4	3–17	9.75	0–17	7.00	1.39
	Taichung	2	4–13	8.50	13–40	12.80	0.66
Kaski: 950–1090 m	Masuli	1	2–5	4.00	5–12	7.75	0.52
Kathmandu: 1360 m	Taichung	1	3–20	8.80	5–12	6.60	1.33
Lalitpur: 1360 m	Taichung	4	0–5	0.67	0–13	3.40	0.67
Makawanpur: 350 m	Himali	1	8–16	12.60	0–4	2.00	6.30
	Masuli	4	3–19	8.70	0–9	5.36	1.62
	Radha–9	1	15–23	19.80	2–9	6.00	3.30
	Taichung	1	5–13	9.33	1–3	1.80	5.18
Morang: 220 m	Masuli	1	99–113	109.00	2–5	3.20	34.06
Nawalparasi: 220 m	Masuli	1	8–14	11.40	2–6	4.00	2.85
Saptari: 250 m	Masuli	1	61–102	83.80	2–5	4.60	18.22
Sarlahi: 250 m	Masuli	1	0–2	1.10	0–2	0.40	2.75
Siraha: 250 m	Masuli	1	78–133	103.60	1–4	2.60	39.85
Sunsari: 200 m	Masuli	1	78–103	91.60	2–5	3.20	28.62
Syanja: 800 m	Masuli	2	3–9	6.00	3–45	16.00	0.36
Tanahu: 400–450 m	Masuli	1	8–17	11.40	3–6	4.20	2.71
	Sabitri	1	7–11	9.00	2–6	3.80	2.37
Local cultivars							
Bara	Unknown	1	0–2	1.20	0–2	0.40	3.00
Chitwan	Acchame	1	3–7	5.20	1–3	1.80	2.89
	Amjhutte	1	1–3	2.00	0–2	0.60	3.33
	Anati	1	1–3	2.20	0–2	0.80	2.75
	Basmati	1	2–7	4.60	1–4	2.40	1.92

Table 5 continued

| Districts: altitude | Rice cultivar | No. of fields | Leafhoppers (no./30 strokes) | | | | Ratio* |
| | | | N. virescens | | N. nigropictus | | |
			Range	Average	Range	Average	
Dhading	Ekka	1	5–10	7.80	12–15	13.60	0.57
	Jethobudho	1	2–12	7.20	3–9	5.00	1.44
	Manabhog	1	2–13	6.40	0–6	2.40	2.67
	Rajbhog	1	5–13	7.20	1–6	3.00	2.40
	Satthia	2	8–37	19.60	3–21	8.20	2.39
	Unknown	1	6–9	7.20	2–5	3.60	2.00
Gorkha	Unknown	1	1–5	2.60	6–11	8.16	0.32
Janakpur	Basmati	1	0–1	0.10	0–1	0.20	0.50
Jhapa	Binas	1	37–51	42.00	1–2	1.40	24.00
Kabhre	Pokhareli	2	11–16	13.00	3–7	4.80	2.71
Kaski	Biringful	1	4–11	6.80	12–15	22.80	0.30
	Darmali	1	2–10	4.40	17–48	28.00	0.16
	Dhima	1	7–12	9.40	8–28	16.80	0.56
	Gorkhali	1	3–8	5.80	14–20	17.00	0.34
	Marsi	1	4–12	6.20	5–15	12.00	0.52
Kathmandu	Pokhareli	1	0–2	0.80	3–7	5.00	0.16
Lalitpur	Marsi	1	0–2	1.80	2–8	4.40	0.41
Makawanpur	Pokhareli	2	3–12	6.50	8–16	13.75	0.42
Syanja	Goladhan	1	3–6	4.20	4–8	5.00	0.84
Tanahu	Acchame	1	4–12	7.80	9–16	12.00	0.65
	Amjhutte	1	1–6	3.20	1–4	2.40	1.33
	Madise	1	3–6	4.20	1–4	2.40	1.75

*Ratio computed as average population of N. virescens/N. nigropictus.
†No insects collected or not applicable.

Population development

The population development of *Nephotettix* spp. was studied under lowland and upland conditions in Chitwan District, a central inner Terai region of Nepal. The leafhopper populations were monitored by net-sweepings at nine locations at weekly or fortnightly intervals during 1992.

The populations of leafhoppers varied depending on cultivar and survey period; the population of *N. virescens* was much higher than that of *N. nigropictus*. The ratio of the two species varied depending on cultivar and period of collection. Although the average numbers of leafhoppers caught increased with the growth stage of the crop, the ratio of *N. virescens* to *N. nigropictus* varied without any clear trend. Leafhopper populations were generally higher on improved than on the local cultivars. On the improved cultivar 'Masuli', populations were low until mid-September and increased thereafter until mid-October. The increase was rapid to reach more than 35 insects/30 sweeps at Rampur, Gunjnagar and Saradanagar, while it was slower in other locations. The ratio of *N. virescens* to *N. nigropictus* was generally below 12, but variable and ranged from 0.6 at Manahari to 11.5 at Rampur (lowland).

Rice green leafhoppers were monitored weekly under contrasting upland and lowland environments at three sites of Chitwan. Leafhopper populations were higher in lowland than upland conditions. Under lowland conditions, the population was higher in seed-beds during March (23–63 insects/30 sweeps) than in the early rice fields during April–May. The leafhopper populations increased in seed-beds and transplanted crops of regular season rice. The populations of *N. virescens* were always higher than those of *N. nigropictus*, but the ratio varied under both lowland and upland conditions. It was generally low in early season rice and higher in main season rice. Studies are in progress to characterize further these populations based on their biology, feeding behaviour and host reactions.

FUTURE RESEARCH STRATEGIES

Dahal (1984) suggested future research needs for rice virus diseases in Nepal as detailed studies on (a) aetiology, (b) host-range, (c) transmission by different vector species, (d) physical and chemical characteristics of the causal viruses, (e) virus–vector–host interactions, (f) screening of rice germplasm against virus diseases and identifying sources of resistance, (g) general and specific control measures, and (h) strengthening collaborative studies with other international organizations. Such studies are currently in progress.

Future research on vector leafhoppers is to be focused on:

* determining population density and species composition from representative ecological zones of the country
* characterization of leafhopper populations based on biological (feeding behaviour, response and development, transmission efficiency), morphological and biochemical (DNA fingerprinting, esterase pattern) characteristics.

ACKNOWLEDGEMENTS

This study was supported by the Rockefeller Foundation, New York, USA (grant no: RF 91003 # 119) and the Institute of Agriculture and Animal Sciences, Rampur, Chitwan, Nepal. Technical assistance of Mr Ram Babu Shrestha is gratefully acknowledged.

REFERENCES

ADHIKARI, R.R., SHRESTHA, G. L. and THAKUR, P. (1986) Review of entomological research in Nepal. pp. 241–278. In: *Thirteenth Summer Crops Workshop, Zanakpur Agricultural Development Program, Naktajhij, Nepal, 10–13 March 1986.*

ANJANEYULU, A., SHUKLA, V.D., RAO, G. M. and SINGH, S.K. (1982) Experimental host range of rice tungro virus and its vector. *Plant Disease,* **66**: 54–56.

AOTA, S. (1979) *Field Problems in Hardinath Farm.* Japan International Co-operation Agency/ Hardinath Agricultural Farm, Hardinath, Jankpur, Nepal.

BAJET, N.B., DAQUIOAG, R.D. and HIBINO, H. (1985) Enzyme-linked immunosorbent assay to diagnose tungro. *Journal of Plant Protection in the Tropics,* **2**: 125–129.

DAHAL, G. (1984) Status of rice virus diseases in Nepal: a review. *Nepalese Journal of Agriculture,* **15**: 173–182.

DAHAL, G. and SHRESTHA, R.B. (1994) *Bibliography on Rice Virus Diseases and Leafhopper and Planthopper Vectors in Nepal.* Rampur, Chitwan, Nepal: Institute of Agriculture and Animal Sciences.

DAHAL, G., AMATYA, P. and MANANDHAR, H.K. (1992) Plant diseases in Nepal. *Review of Plant Pathology,* **72**: 797–807.

DAHAL, G., SHRESTHA, R.B., KHATRI, N.K., FAN, Z. and HULL, R. (1993) Incidence of virus diseases of rice in Nepal. *Journal of the Institute of Agriculture and Animal Sciences,* **14**: 115–116.

GARRITY, D.P., SHEPPARD, D.R., MINNIK, D.R. and BONMAN, J.M. (1987) Accelerating rice research and production in Nepal. *Report of an IRRI Team Mission, 28 August–9 September 1987.* Los Baños, Philippines: International Rice Research Institute.

HIBINO, H., ROECHAN, M. and SUDERISMAN, S. (1978) Association of two types of virus particles with Penyakit Habang (tungro disease) of rice in Indonesia. *Phytopathology,* **68**: 1412–1416.

INTERNATIONAL RICE RESEARCH INSTITUTE (1983) *Annual Report for 1982.* Los Baños, Philippines: International Rice Research Institute.

INTERNATIONAL RICE RESEARCH INSTITUTE (1990) *World Rice Statistics.* Los Baños, Philippines: International Rice Research Institute

INOUE, H. (1986) Biosystematic study on the genus *Nephotettix* occurring in Asia. *Bulletin of Kyushu Agricultural Experimental Station,* **24**: 150–255.

JOHN, V.T., FREEMAN, W.H. and SHAHI, B.B. (1979) Occurrence of tungro in Nepal. *International Rice Research Newsletter,* **4**: 16.

JOSHI, S.L. (1972) The progress report of paddy. pp. 14–16. In: *Proceedings of the Summer Crops Seminar, 1972, Parwanipur, Bara, Nepal.*

JYOTI, J.L. (1989) Studies on native pests of rice and their prevalence in relation to rice fields in Nepal. pp. 213–224. In: *Rice Improvement Programme Reports, the Fourteenth Summer Crops Workshop, 17–25 January 1989,* National Rice Improvement Programme, Parwanipur, Bara, Nepal.

JYOTI, J.L. (1990) Evaluation of eight insecticides in controlling rice insect pests. pp. 171–175. In: *Proceedings of the Fourth Summer Crops Workshop and Programme Planning, 4–9 February 1990,* National Rice Improvement Programme, Parwanipur, Bara, Nepal.

LING, K.C. (1972) *Rice Virus Diseases.* International Rice Research Institute, Los Baños, Philippines.

LING, K.C. and TIONGCO, E.R. (1980) *Rice Virus Diseases in the Philippines.* Los Baños, Philippines: International Rice Research Institute.

MALLIK, R.N. (1981) *Rice in Nepal.* Kala Prakashan, Nepal.

OMURA, T., INOUE, H., THAPA, U. B. and SAITO, Y. (1981) Association of rice tungro spherical and rice tungro bacilliform virus with the disease in Janakpur, Nepal. *International Rice Research Newsletter,* **6**: 14.

OMURA, T, HIBINO, H., USUGI, T., INOUE, H., MORINAKA, T., TSRUMACHI, S., ONG, C.A., PUTTA, M., TSUCHIZAKI, T. and SAITO, Y. (1984) Detection of rice viruses in plants and individual insect vectors by latex flocculation test. *Plant Disease,* **68**: 374–378.

OU, S. H. (1985) *Rice Diseases.* 2nd edn. Kew, UK: Commonwealth Mycological Institute.

PALIKHE, B.R. (1982) Studies on occurrence and population fluctuations of insect pests by light trap. *Proceedings of the Eighth Summer Crops Workshop, 25–30 January 1982,* National Maize Development Programme, Rampur Agricultural Research Station, Rampur, Chitwan, Nepal.

PRADHAN, R.B. (1973) Studies on the biology of *Nephotettix apicalis* under laboratory and field conditions. *Annual Report of the Entomology Division for 1972/73.* Khumaltar, Lalitpur, Nepal: Department of Agriculture.

PRADHAN, R.B. (1979) Studies on the population dynamics of insect pests by light trap method. *The Seventh Rice Improvement Workshop, 24–27 February 1980.* National Rice Improvement Programme, Parwanipur, Bara, Nepal.

PRADHAN, R.B. (1980) Study on *Nephotettix apicalis* under laboratory and field conditions. pp. 231–237. In: *Annual Report of the Entomology Division for 1972–73.* Khumaltar, Lalitpur, Nepal: Department of Agriculture.

PRADHAN, R.B. (1983) Report on the incidence of rice pests in 1982. *The Tenth Summer Crops Workshop, 23–28 January 1983,* National Maize Development Programme, Rampur, Chitwan, Nepal.

PRADHAN, R.B. (1989) Review of entomological works on paddy in Nepal. *Journal of Entomological Society of Nepal,* **1**: 91–102.

PRADHAN, S.B. and JOSHI, S.L. (1972) Biological study of *Nephotettix apicalis* Motsch. in rice in laboratory and field conditions. pp. 14–16. In: *Progress Report on the Paddy.* Khumaltar, Lalitpur, Nepal: Department of Agriculture.

PRADHAN, R.B. and KHATRI, N.K. (1980) The varietal performance of some popular varieties of rice against white-back planthopper (*Sogatella furcifera*) and green rice leafhoppers (*Nephotettix apicalis*). *The Seventh Rice Improvement Workshop, 25–30 February 1981,* Rampur, Chitwan, Nepal: Rampur Agricultural Station.

RAO, G.M. and ANJANEYULU, A. (1978) Host range of rice tungro virus. *Plant Disease Reporter,* **58**: 856–860.

RIVERA, C.T. and OU, S.H. (1965) Leafhopper transmission of tungro disease of rice. *Plant Disease Reporter,* **49**: 127–131.

RIVERA, C.T., LING, K.C. and OU, S.H. (1969) Host range of rice tungro virus. *Philippine Phytopathology,* **6**: 16–17.

SHARMA, K.C. and KHATRI, N.K. (1977) Studies on the population dynamics and control of paddy hoppers in the field conditions. pp. 57–62. In: *The Fifth Rice Improvement Workshop, 27 February– 2 March 1978.* National Rice Improvement Programme, Parwanipur, Bara, Nepal.

SHARMA, K.C. and MATHEMA, S.R. (1976) Studies on field biology and control of leafhoppers associated with paddy. pp. 33–34. In: *The Fourth Rice Improvement Workshop, 15–18 March 1977.* National Rice Improvement Programme, Parwanipur, Bara, Nepal.

STHAPIT, B.R. (1992) Chilling injury of rice crops in Nepal: a review. *Journal of the Institute of Agriculture and Animal Sciences,* **13**: 1–32.

WATHANAKUL, L. (1964) *A Study of Host Range of Tungro and Orange Leaf Virus of Rice.* MSc Thesis, University of the Philippines, College of Agriculture.

Paper 12

Rice tungro virus disease in Indonesia: present status and current management strategy

A. HASANUDDIN, KOESNANG AND D. BACO

Maros Research Institute, PO Box 173, Ujung Pandang, South Sulawesi, Indonesia

INTRODUCTION

Indonesia is one of the largest nations in the tropics with a population exceeding 180 million. More than 9 or 10 million ha of rice are grown every year with most intensive production in Java, Bali and the southern part of Sulawesi (Ruchiyat and Sukmaraganda, 1991). During the last 25 years much effort has been made to increase rice production to meet the demand created by an expanding population which increases annually by *c.* 2.3%, necessitating about half a million tonnes of additional food. Although self-sufficiency in rice has been achieved since 1984, further research is necessary to help achieve stability in production and to ensure further improvement in crop yield potential. The main problem in stabilizing and increasing rice production in Indonesia is the continuous threat of pests and diseases. Rice tungro virus disease – commonly referred to as rice tungro disease or tungro – which is transmitted by leafhopper vectors, is one of the most important diseases. In general, losses due to the disease vary from year to year, depending on varietal susceptibility, the control measures adopted and environmental factors. Between 1969 and 1992 there were serious outbreaks in various regions (Table 1) and 23 of the 27 provinces of Indonesia were affected (Figure 1). It is likely that such problems will continue unless effective control measures are implemented, particularly in the endemic areas.

Outbreaks of tungro disease are always associated with the population build-up of the most important vector, the green leafhopper (GLH) *Nephotettix virescens* (Distant), as other vector species are relatively unimportant (Siwi, 1986). In tungro outbreak areas in Indonesia, such as Central and East Java and South, South-east and Central Sulawesi, 99% of the individuals of all *Nephotettix* spp. recorded were *N. virescens* — *N. nigropictus* or other *Nephotettix* spp. were not found, even in weeds surrounding rice fields. The species composition of *Nephotettix* populations in the field plays an important role in the incidence of tungro disease.

This paper discusses the present status of tungro in Indonesia and the research activities of the Maros Research Institute for Food Crops (MORIF) in developing a management strategy to curb the problem.

Table 1   Incidence of rice tungro disease in Indonesia, 1962–92

Period	Region	Area infected (ha)	Variety affected
1969–72	Sumatra	21 000	Kwatik
	Kalimantan	5 000	Lemo
1973–75	Sulawesi	97 254	IR5 Pelita Local
1976–83	Sulawesi	22 721	IR20 IR26 IR36 IR42 Cisadane
1980–83	Bali	16 000	IR36 IR42 IR50 IR54
1981–83	Java	12 316	Cisadane IR36 IR42
	Nusatenggara	12 091	IR36 IR42
1982–83	Sumatra	4 109	IR26 IR36 IR42
	Kalimantan	8 015	IR36 IR42
	Maluku	106	IR36
	Irian Jaya	46	IR36 IR42
1984–92	Sulawesi	18 985	IR54 IR64
	Bali	16 481	IR36 IR54 K Aceh
	Nusatenggara	10 652	IR54 IR64
	Java	11 837	Cisadane IR36 K Aceh
TOTAL		244 904	

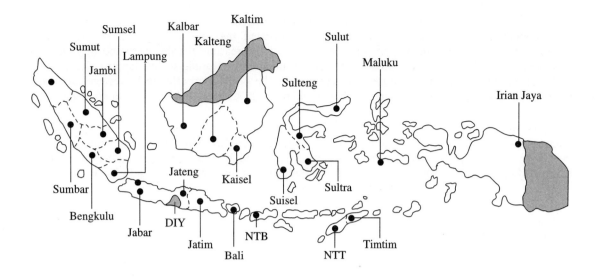

Figure 1 Location of outbreaks of rice tungro disease in Indonesia, 1969–92 (Anon., 1992).

HISTORICAL PROFILE OF TUNGRO INCIDENCE

Van der Vecht (1953) reported *mentek* disease of rice which caused yellow or reddish-brown discolouration and stunted growth in several Indonesian islands. This disease was similar to the one later described as 'tungro' (Ou, 1965; Rivera *et al.*, 1968). It is called *cella pance* in South Sulawesi, *penyakit habang* in South Kalimantan, *kebebeng* in Bali and *mentek* in Java.

Tungro disease was not considered a major constraint to rice production in the 1950s and 1960s because of the widespread use of resistant varieties. However, in the early 1970s tungro outbreaks were reported from South Kalimantan, South Sumatra, Lampung and South and Central Sulawesi where 50 000 ha of rice were damaged (Oka, 1971; van Halteren and Sama, 1973). Manwan *et al.* (1985) estimated that between 1968 and 1984 the total area of Indonesia affected by tungro disease was 199 000 ha. The area affected by tungro later declined gradually and has since remained relatively low (Figure 2).

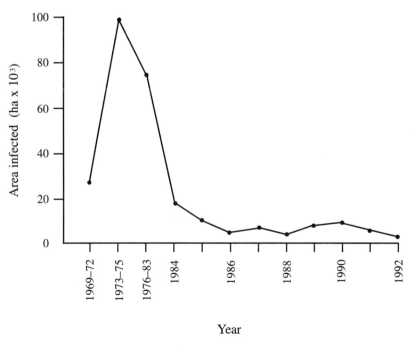

Figure 2 Profile of tungro incidence in Indonesia, 1969–92.

Tungro gained importance after the introduction of new high-yielding rice varieties, most of which were susceptible to the disease. Ling *et al.* (1983) stated that the development of high-yielding varieties and changes in cultural practices were two of the reasons for the increased importance of *Nephotettix* spp. High-yielding varieties were grown intensively, leading to overlapping rice crops. In such circumstances, the tungro viruses may survive and be propagated effectively because susceptible varieties are good sources of inoculum both as standing crops and as stubble regrowth after harvest.

In 1969 the Government of Indonesia (GOI) introduced and distributed varieties bred by the International Rice Research Institute (IRRI), including IR5 and IR8 which became popular because of their high yield potential and resistance to bacterial leaf blight. The improved Indonesian varieties Pelita I-1 and Pelita I-2 were released in 1971. Pelita I-1 became popular because of its high yield, resistance to bacterial leaf blight, good grain characteristics and very good eating quality.

Following the tungro outbreaks in 1969–83, GOI introduced several resistant varieties including C4-63, IR20, IR26, IR30 and IR34. These varieties were increasingly affected by tungro with time, due to the adaptation of the leafhoppers and the development of new biotypes able to thrive on previously resistant varieties (Sama *et al.*, 1983; Manwan *et al.*, 1985). The same phenomena also occurred in Bali and Central and East Java in 1980–81 (Tantera, 1982), when previously resistant varieties including IR36, IR38, IR42, IR50, IR52 and IR54 were seriously damaged by tungro. Recommendations to farmers for tungro management during 1972–81 were:

- cultivation of resistant high-yielding varieties
- regular monitoring of GLH populations and the application of appropriate insecticides if numbers reached two or three per hill and a source of inoculum was present
- removal of infected plants showing tungro symptoms, especially during the early stage of infection
- ploughing-in of stubbles and weeds immediately after harvest to eliminate sources of inoculum
- avoidance of overlapping rice crops which would otherwise enhance the population build-up of the vector and increase tungro incidence.

An integrated tungro management scheme was developed by MORIF in 1982 and introduced in South Sulawesi (Sama *et al.*, 1991) where tungro incidence has since declined (Figure 3). Consequently, the use of insecticides for pest management was greatly reduced and populations of the brown planthopper (BPH), *Nilaparvata lugens*, and stemborers also declined. It is considered that the integrated tungro management scheme has prevented the adaptation of GLH to some varieties which have maintained their vector resistance with time (Table 2).

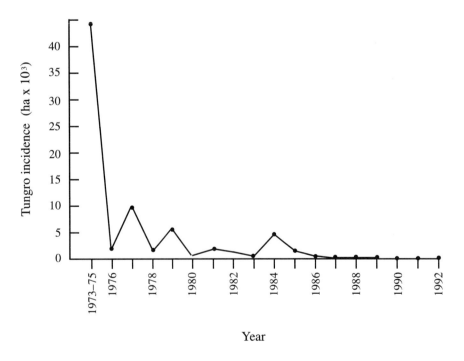

Figure 3   Profile of tungro incidence in South Sulawesi, 1973–92 (Baco, 1993).

96

Table 2 Reaction of rice varieties to field colonies of GLH tested at different times in Maros from 1989 to 1992 using the 'wooden box' test method and the IRRI Standard Evaluation System for rice

Variety	1989	1990	1991	1992
Pelita	MS	MS	MR	R
Serayu	MR	MR	MR	R
Berantas	R	MR	R	R
IR26	MR	MR	R	R
IR42	R	MR	R	R
IR48	R	MR	R	R
IR54	R	MR	R	R
IR64	MR	MR	R	R
IR72	R	MR	R	R

R = resistant, MR = moderately resistant; MS = moderately susceptible; s = susceptible.

DEVELOPMENT OF A MANAGEMENT STRATEGY TO CONTROL TUNGRO

Of the many possible methods of controlling tungro disease, the most practical at present are the use of resistant varieties, chemical control of the vectors and the manipulation of cultural practices. The use of resistant varieties has always been considered the most effective, simplest and most economical means of control which avoids problems of environmental pollution. However, several GLH-resistant varieties which previously showed field resistance to tungro subsequently became infected when grown continuously over large areas (Manwan *et al.*, 1985).

After many years of experimental work and careful observation in the field, entomologists and plant pathologists at MORIF developed an integrated tungro management strategy with three components: synchronous planting at appropriate planting times in wet and dry seasons; varietal rotation based on the deployment of the GLH-resistance genes available; and judicious use of insecticides. Utilization of these components for tungro management reduced the yield losses due to tungro.

Synchronous planting at appropriate planting dates

Results showed that populations of GLH and tungro incidence fluctuated in rice fields from season to season; contributing factors are varietal susceptibility, populations of natural enemies and climatic conditions (Manwan *et al.*, 1985). Several studies between 1977 and 1990 in South Sulawesi indicated that GLH numbers recorded in light traps had a major peak near the end of the wet season and a smaller peak near the end of the dry season crop (Figure 4). The GLH population declined after harvest in both seasons and was lowest during fallow periods.

Monthly planting trials revealed that mean tungro incidence was high in plots planted when GLH numbers were also high. This occurred in August–September on the east coast (Maros) and in February–March on the west coast (Sidrap, in Langrang regency). The optimum time to plant rice at Maros to reduce the risk of tungro is December–January in the wet season and June–July in the dry season. In Langrang, the recommended planting time is from April to May in the wet season, whereas in areas between the east and west coast regions the optimum time is in December or January. Plantings at the recommended times reach the booting stage (60–70 days old) before high GLH numbers occur, by which time plants have become less susceptible to tungro infection.

Varietal rotation and resistant varieties

Collaborative research between the Agency for Agricultural Research and Development (AARD), MORIF and IRRI was begun in 1986 to determine the efficiency of transmission of rice tungro viruses by colonies of *N. virescens,* previously maintained on IR54 and TN1, to selected 'IR' varieties of different genetic background. Both IR54 and TN1 source plants, infected with both rice tungro bacilliform virus (RTBV) and rice tungro spherical virus (RTSV), were used in the transmission studies. Tungro incidence was assessed by a visual score. Regardless of the disease source, transmission by the IR54 colony on varieties IR50 and IR54 was much higher than by the TN1 colony (Figure 5). This was confirmed when the plants were indexed serologically for RTBV and RTSV using the latex test. These findings showed that the colony reared continuously on IR54 had already become adapted to this variety.

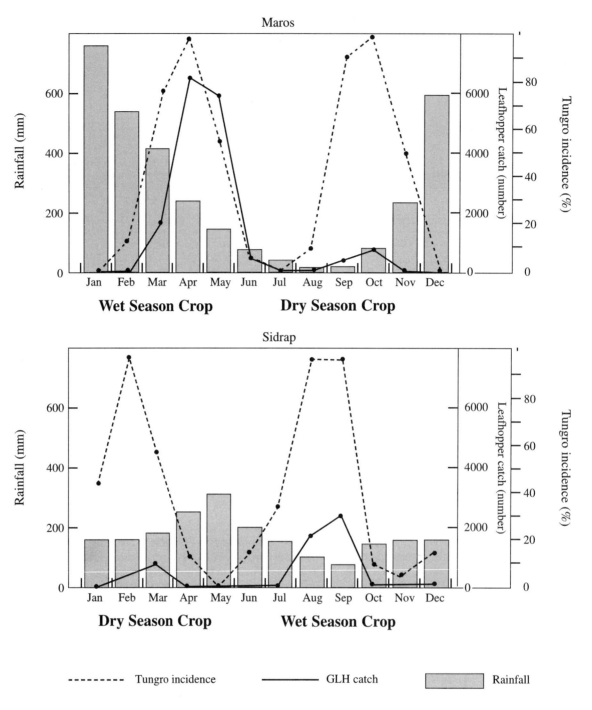

Figure 4   Green leafhopper (GLH) catches in light traps and cropping patterns for wet and dry seasons, average monthly rainfall for five years and average tungro incidence in Pelita I-1 planted each month during 1977–81 in Maros and Sidrap.

Manwan *et al.* (1985) reported that varietal reaction to GLH differed between locations. The differential reaction of several rice varieties to colonies of *N. virescens* under field and greenhouse conditions is shown in Table 3. These data indicate that the reaction of resistant varieties to tungro infection varied between locations and between years. Further evidence of a changed response is provided by IR36 which still showed resistance to tungro infection in South Sulawesi in 1980, when it was severely affected by tungro in Bali.

Seven genes conferring resistance to GLH have been reported and designated as GLH-1 to GLH-7 (International Rice Research Institute, 1983). In order to prevent the formation of GLH colonies on resistant varieties in the field, varietal rotation based on GLH resistance is needed. Four groups of rice

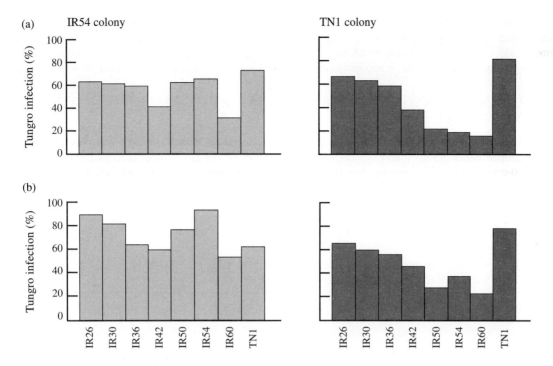

Figure 5   Tungro infection based on symptoms of rice varieties inoculated in test tubes by IR54 and TN1 colonies of GLH at 1 insect/seedling using plants of (a) Pelita and (b) IR54 infected with both RTBV and RTSV as inoculum source (Amran *et al.*, 1989).

varieties were classified at Maros, based on their reaction to tungro and designated as T_0, T_1, T_2 and T_3 (Manwan *et al.*, 1985). Under the integrated tungro management scheme it was recommended that each group of varieties should be deployed in an appropriate rotation cycle.

Table 3   Reaction of rice varieties to tungro virus tested at four locations in Sulawesi in 1982–93 using the 'wooden box' test method and the IRRI Standard Evaluation System for Rice

| Variety* | Test site (location) of GLH colony | | | |
	Lanrang	Maros	Mariri	Parigi
Pelita	S	S	S	S
Cisadane	S	S	S	S
IR30	R	R	R	R
IR42	S	R	S	S
IR54	S	R	R	R
Kelara	R	R	R	R
Bahbolon	–	R	R	R

*Most commonly planted varieties around the test site were:
  Lanrang (South Sulawesi)   IR36   IR42   IR54   IR56
  Maros (South Sulawesi)   Cisadane
  Mariri (South Sulawesi)   IR36   IR42
  Parigi (Central Sulawesi)   IR36   IR42
  R = resistant; S = susceptible; – = not tested.

Judicious use of insecticides

There are two approaches to chemical control of vectors in Indonesia: to use chemicals to eradicate the vectors in infected fields to prevent them dispersing to neighbouring fields; and to protect healthy crops from infection. However, the majority of farmers will only apply insecticides when they see obvious damage due to pests or diseases. The decision as to whether to apply insecticides depends very much on the farmers' experience and to some extent on the advice from the surveillance and monitoring network.

Results from trials at MORIF showed that some insecticide applications provided good control of GLH and helped in reducing tungro incidence. However, results have also shown that repeated use of the same insecticide caused it to become ineffective in controlling tungro, especially under conditions of high vector and inoculum pressure. It appears that GLH has developed a degree of resistance to insecticides.

CURRENT TUNGRO MANAGEMENT RECOMMENDATIONS AND ACTUAL FARMERS' PRACTICES

The integrated tungro management scheme which was developed by MORIF in 1983 is now being practised by farmers in South Sulawesi and is recommended for other provinces. In order to implement this concept close co-operation is needed among the various government agencies concerned and researchers, extension workers and farmers for proper planning to determine appropriate planting times and the varietal rotation to be followed each growing season. An annual meeting between MORIF researchers, extension specialists, University staff and members of the mass guidance rice programme technical team is held to develop control recommendations for the next full year of rice cropping (Table 4).

Table 4 Control recommendations for the rice crop year (1993/1994) in South Sulawesi in three different tungro management regions

Activities	East Coast		West Coast		Between East and West Coast	
	Wet	Dry	Wet	Dry	Wet	Dry
1. Cropping system	Rice-Rice-Palawija Rice-Palawija/Horticulture		Rice/Minapadi-Palawija		Rice-Rice-Palawija Rice-Rice-Fallow	
2. Planting time	Dec–Jan	May–Jun	Apr–May	Nov–Dec	Dec–Feb	Jun–August
3. Varieties	Cisadane	IR72	IR66	PB42	PB42	Ciliwung
	Atomita-4	IR64	Ciliwung	IR64	IR74	IR48
	PB42	IR66	IR70	Seililin	IR64	IR70
	Ciliwung	IR74	IR72	IR74	Seililin	IR72
	IR70	Seililin	IR42	–	Atomita-4	IR66
4. Dosage of fertilizer (kg/ha)						
Urea	250	200	150–200	200–250	200	150
TSP	50	25–50	25–50	25–50	100	50–100
KC1	50	50	50–100	50–100	100	100
ZA	50	50	50	50	50	50
	Based on recommendation of Directorate General Agriculture of Food Crop					
5. Pests and diseases (expected)	Rodent	Rodent	Rodent	Rodent	Rodent	Rodent
	BPH	Rice bug	BPH	Caseworm	Cutworm	BPH
	Caseworm		Rice bug	and Blast	BLB	Cutworm
	BLB, Blast		Caseworm		and Blast	Rice bug
	and Cutworm					
6. Pesticide (depends on economic threshold and natural enemies)	Carbofuran (17 kg/ha)	Carbofuran (17 kg/ha)	Klerat RHB (2 kg/ha), Carbofuran (34 kg/ha)	Klerat RHB (1 kg/ha), Carbofuran (17 kg/ha)	Klerat RMB Carbofuran (17 kg/ha)	Klerat RMB Carbofuran (17 kg/ha)
7. To prevent rodents	Bumbu rat	Bumbu rat	Bumbu rat	Bumbu rat	Bumbu rat	Bumbu rat

BPH = *Nilaparvata lugens*
BLB = *Xanthomonas campestris* pv *oryzae*
Rice bug = *Leptocorisa oratorius*
Blast = *Pyricularia oryzae*
Caseworm = *Nymphula depunctalis*
Cutworm = *Spodoptera litura*

Table 5 Tungro survey for South Sulawesi (MORIF, 1993)

Control measure/locality	Current recommendations		Basis for recommendations	Farmer adoption		
	Yes	No		Often	Sometimes	Never
Planting date	✓		Monitoring vector population	✓		
Synchronize planting	✓		Different duration of maturity of each variety	✓		
Fallow period	✓		To eliminate ratoons and weeds		✓	
Direct seeding		✓				
Early season vector control		✓				
Vector control during outbreaks	✓		To suppress vector populations	✓		
Seed-bed protection		✓				
Roguing						
Extra nitrogen	✓		To enhance recovery	✓		
Destruction of infected crop and ploughing stubbles	✓		To reduce source of inoculum	✓		
Rotation varieties	✓		To maintain degree of resistance	✓		

✓ Actual farmers' practices. No other measures adopted.

Although the management scheme appears to be effective in reducing tungro incidence, its implementation requires effective co-ordination and planning and a well-organized seed production system to ensure that seed of recommended varieties is available. The management scheme can be modified further to facilitate dissemination of this control strategy. Further recommendations, and the level of farmer adoption, are listed among a range of potential tungro control measures in Table 5.

RESEARCH AND TRAINING NEEDS

There is a need to develop training in tungro control methods and to assess the scope for future collaboration to improve the research capability. The training programme and future collaborative research should cover the following areas:

- training on identification of tungro disease and leafhopper vectors and on appropriate control methods
- studies on tungro disease management in relation to epidemiology, vector populations, rainfall pattern and the role of alternative hosts of GLH
- research on the use of botanical insecticides and antifeedants, which disturb GLH
- studies on the genetic variability in vector populations and monitoring of shifts in their ability to infect rice varieties over time
- an examination of the factors contributing to RTVD endemicity and to 'hot spots', so that outbreaks can be predicted.

To achieve these objectives, it is essential to ensure close international collaboration with adequate long-term funding for joint research projects, regular seminars and workshop training courses and exchanges of scientists.

REFERENCES

AMRAN, M., SUDJAK, M., BASTIAN, A., HASANUDDIN, A., CABUNAGAN, R.C. and HIBINO, H. (1989) Transmission of rice tungro-associated viruses by green leafhopper (*Nephotettix virescens* Distant) after selection on IR54 rice variety. *Agrikam*, **4**(2): 55–62.

ANON. (1992) Tungro dan Wereng hijau, Ditlin Tanaman Pangan, Dirjen Tanaman Pangan.

BACO, D. (1993) Integrated pest management on rice in south Sulawesi. Status and Research Needs. Paper presented at *Rice IPM Network Thailand–China Study Visit. Workshop held at Bangkok, 18–31 May 1993.*

VAN HALTEREN, P. and SAMA, S. (1973) The tungro disease in South Sulawesi. Paper presented at Staff Meeting, 25–28 July 1973, Bogor, Indonesia. [Mimeo.]

INTERNATIONAL RICE RESEARCH INSTITUTE (1983) *Annual Report 1982.* pp. 62–65. Los Baños, Philippines: International Rice Research Institute.

LING, K.C., TIONGCO, E.R. and FLORES, Z.M. (1983) Epidemiological studies of rice tungro. pp. 249–257. In: *Plant Virus Epidemiology. The Spread and Control of Insect-Borne Viruses.* PLUMB, R.T. and THRESH, J.M. (eds). Oxford: Blackwell Scientific Publications.

MANWAN, I., SAMA, S. and RISVI, S.A. (1985) Use of varietal rotation in the management of tungro disease in Indonesia. *Indonesian Agricultural Research and Development Journal,* **7**(3,4): 43–48.

OKA, I.N. (1971) An outbreak of rice disease showing tungro symptoms in South Sumatra and Lampung provinces. Ministry of Agricultural Research InsAgr, Bogor, Indonesia. [Mimeo.]

OU, S.H. (1965) Rice diseases of obscure nature in tropical Asia with special reference to '*mentek*' disease in Indonesia. *International Rice Commission Newsletter,* **14**(2): 4–10.

RIVERA, C.T., OU., S.H. and TANTERA, D.M. (1968) Tungro disease of rice in Indonesia. *Plant Disease Reporter,* **52**: 122–124.

RUCHIYAT, E. and SUKMARAGANDA, T. (1991) National integrated pest management in Indonesia: its successes and challenges. pp. 329–347. In: *Proceedings of the Conference on IPM in the Asia–Pacific Region.* Kuala Lumpur, Malaysia: CAB International/Asian Development Bank.

SAMA, S., HASANUDDIN, A. and SUPRIHATNO, B. (1983) Penelitian penyakit tungro dan wereng hijau. pp. 111–132. *Masalah dan hasil penelitian padi.* Pusat Penelitian dan Pengembangan Tanaman Pangan, Bogor.

SAMA, S., HASANUDDIN, A., MANWAN, I., CABUNGAN, R. and HIBINO, H. (1991) Integrated management of rice tungro disease in South Sulawesi, Indonesia. *Crop Protection,* **10**: 35–40.

SIWI, S.S. (1986) On the occurrence, morphology and biology of *Nephotettix* species in Indonesia. *Proceedings of the Second International Workshop on Leafhoppers and Planthoppers of Economic Importance.* Provo, Utah, USA.

TANTERA, D.M. (1982) Serangan penyakit tungro di bali. *Journal Litbang Pertanian,* **1**: 2–5.

VAN DER VECHT, J. (1953) The problem on the *mentek* disease of rice in Java. Contrib. Gen. Agricultural Research Station. No. 1–37. Bogor, Indonesia: Agricultural Research Station.

Incidence and control of rice tungro disease in Malaysia

Y.M. CHEN and A.T. JATIL

Department of Agriculture, Jalan Gallagher, 50480 Kuala Lumpur, Malaysia

INTRODUCTION

Rice tungro virus disease, often referred to as tungro, or *penyakit merah* disease as it is known in Malaysia, has caused devastating epidemics in rice in South and South-East Asia. Plants affected by tungro are stunted, show various degrees of discoloration, give poor yields and mature late. The disease is caused by a complex of two viruses – rice tungro spherical virus (RTSV) and rice tungro bacilliform virus (RTBV) (Hibino *et al.*, 1978, 1991; Omura *et al.*, 1983; Saito, 1977). RTBV contains circular double-stranded DNA while RTSV contains single-stranded RNA. Both viruses are transmitted in a semi-persistent or transitory manner by the green leafhopper (GLH), *Nephotettix virescens* (Distant), and also by *N. nigropictus* (Stål), *N. malayanus* (Ishihara and Kawase), *N. parvus* (Ishihara and Kawase) and *Recilia dorsalis* (Motschulsky). The different vector species differ considerably in their transmission ability (Ling, 1966). Leafhopper transmission of RTBV depends on the presence of RTSV, while RTSV can be transmitted independently (Hibino *et al.*, 1978, 1979).

OCCURRENCE OF TUNGRO DISEASE

Tungro has been known in Malaysia since 1938 (Ou, 1985). At first it was considered to be a physiological disorder and was later suspected of being caused by nitrogen deficiency (Lockard, 1959), or by hydrogen sulphide or organic acids originating from the anaerobic decomposition of weeds. It was only in the 1960s that the disease was shown to be transmitted by vectors and attributed to a virus (Ou, 1965; Ou and Goh, 1966; Singh, 1969).

Initially, outbreaks of tungro were reported mainly in Krian District, Perak, where a very serious outbreak occurred in 1969 (Lim, 1972). After 1980, however, tungro was reported repeatedly from several other localities in the states of Penang, Kedah and Perlis and in 1982 more than 17 507 ha of paddy were affected by the disease (Department of Agriculture, 1983) (Figure 1a). The yield loss in 1982 was estimated to be RM 22 million (US$ 8.8 million). The worst-affected areas were Perlis and Kedah — both outside the area of the Muda Agricultural Development Authority (MADA) — Krian District, Perak, and especially the Muda Irrigation Scheme (Figure 1b). In 1981, tungro occurred mainly in the southern part of the scheme where 5884 ha were affected; in 1982 the disease severely damaged 5839 ha and reached a maximum of 8655 ha in 1983. The value of crop losses from 1981–83 was estimated to be US$ 10 million.

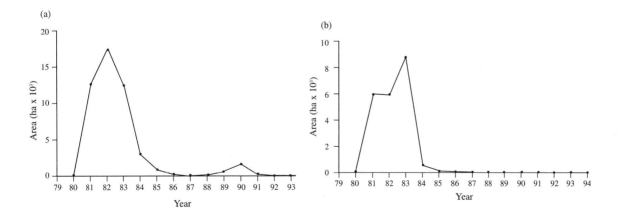

Figure 1 Incidence of tungro in (a) Peninsular Malaysia and (b) the Muda Scheme, 1979–94.

As tungro had such disastrous effects, the Department of Agriculture (DOA) launched a nationwide campaign to control the disease (Department of Agriculture, 1983). This co-operative campaign, undertaken by DOA, the Malaysian Agricultural Research and Development Institute (MARDI), MADA, Kemubu Agricultural Development Authority (KADA) and the Farmers' Organization Authority (FOA) and the Krian-Sungai Manik Project, has greatly reduced its incidence. The area affected by tungro in Peninsular Malaysia was reduced to 12 500 ha in 1983; 3100 ha in 1984 and 1000 ha in 1985 (Figure 1a, Department of Agriculture, unpublished). From 1986 to July 1994 the area affected by tungro was between 330 and 20 ha and the total area of rice grown ranged from 540 000 ha to 730 000 ha.

In 1990 there was a resurgence of tungro disease and 1800 ha were infected sporadically; prompt action was required because the planting of paddy was widely staggered due to problems in water management and because of the high population levels of the main vector, *N. virescens*. An intensive survey was carried out to determine the extent of the disease and infected plants and stubbles were destroyed with insecticide or rogued. Insecticides were used in areas with high vector population. Subsequently, the affected area was greatly reduced to only 67 ha in 1992, 24 ha in 1993 and 20 ha in 1994 (until July).

CONTRIBUTORY FACTORS TO THE 1982 OUTBREAKS

Some of the predisposing factors that contributed to the 1982 outbreaks are listed below:

- **Continuous source of inoculum** The inadequate irrigation system, the limited number of tractors for land preparation and uncertainty in the availability of labour for transplanting and harvesting resulted in excessive staggering of planting dates. Consequently, there was no clear demarcation between the two successive planting seasons. This led to a continuity in the rice habitat available for vectors and hosts of the viruses.

- **Vector population** The high population of active vectors increased the potential for virus transmission; light-trap catches in Telok Chengai, Alor Setar in August 1982 and August 1983 were 247 470 and 149 368, respectively (Figure 2). Since 1980 *N. virescens* accounted for 60–90% of the total number of leafhoppers caught. Previously, more than 60% of the GLH collected from light traps in Muda were *N. nigropictus*, which is a relatively inefficient vector. In Krian District 93% of the leafhoppers collected were *N. virescens* which is far more efficient (Ishihara and Kawase, 1968; Singh, 1971).

- **Susceptible varieties** During and prior to the tungro epidemic the predominant varieties, Serribu Gantang, Anak Dara and MR1, were highly susceptible to GLH and to tungro (Figure 3).

- **Susceptible plant stages** The staggered planting due to water shortage prolonged the period over which rice at susceptible stages of growth was available to virus infection and colonization by vectors. In areas with staggered planting, fields that are planted late have a higher risk of tungro infection because of the increased availability of tungro sources to vectors.

TUNGRO CONTROL PROGRAMME

In response to the epidemics in the early 1980s the various organizations mentioned above co-operated in designing an integrated pest management programme to contain the tungro threat. This three-year programme, '*Campaign to Control Tungro*', had several components.

- **Training** Staff from the DOA and the various extension agencies were trained so that they could teach farmers to recognize tungro symptoms, understand the epidemiology of the disease and carry out appropriate control measures. Three hundred and eighty-nine courses were held in the first quarter of 1983 involving 68 agricultural officers, 628 agricultural technicians and 10 704 farmers (Department of Agriculture, 1983).

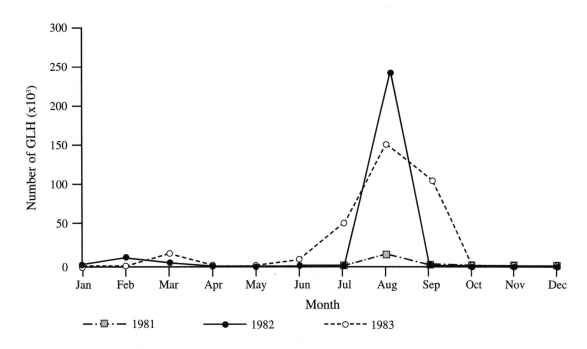

Figure 2 Monthly light-trap catches of green leafhopper (GLH) at Telok Chengai, Alor Setar, August 1981–October 1983.

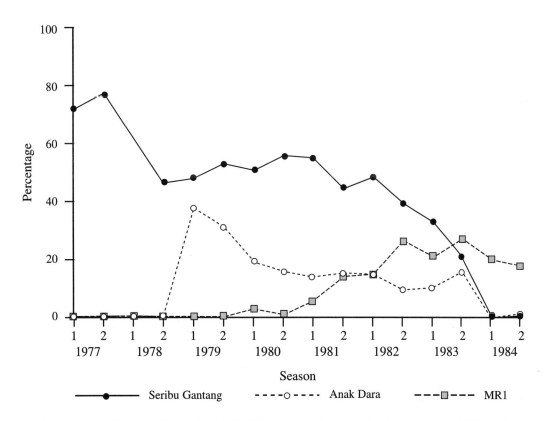

Figure 3 Main paddy varieties in the Muda Scheme from the first planting season of 1977 to the second season of 1984.

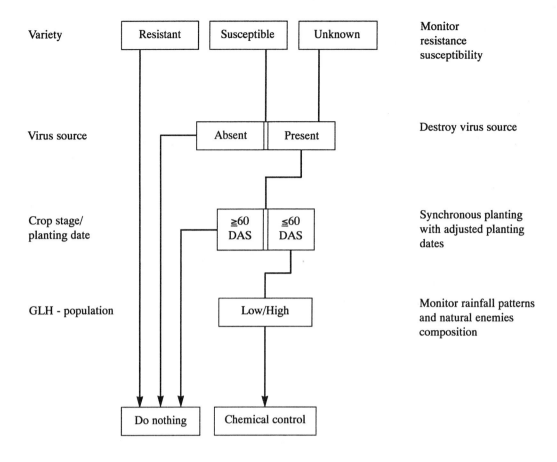

Variety	Resistant	Susceptible	Unknown	Monitor resistance susceptibility
Virus source		Absent	Present	Destroy virus source
Crop stage/ planting date		≥60 DAS	≤60 DAS	Synchronous planting with adjusted planting dates
GLH - population		Low/High		Monitor rainfall patterns and natural enemies composition
	Do nothing	Chemical control		

DAS = Days after sowing   GLH = Green leafhopper

Figure 4   Expert system for rice tungro virus disease.

-

- **Surveillance and forecasting of tungro** Systems for surveillance and forecasting of tungro were set up in all the major production areas, namely Perlis, Kedah, Penang, Perak, Selangor, Kelantan and Terengganu. They involve four main activities:

  - field-scouting to determine the incidence of tungro and to assess GLH populations
  - setting up light traps to detect the build-up of GLH populations and species composition
  - collecting GLH from the field to test their infectivity
  - deployment of mobile nurseries to detect the presence of any viruliferous GLH one month before planting. The approach involved planting TN1 (Taichung Native 1) rice seedlings in trays which were then exposed to GLH in the field. Initially the seedlings were assessed visually for tungro and by using the starch iodine test. They are now tested serologically for each of the tungro viruses by enzyme-linked immunosorbent assay (ELISA).

- **Resistant varieties** In the first season of 1982 the varieties Anak Dara and Seribu Gantang accounted for 15% and 46%, respectively, of all plantings in the MUDA scheme (Perangkaan Padi, 1982). As noted earlier, these two varieties are highly susceptible to the vector and tungro viruses. Accordingly, GLH-resistant or moderately resistant varieties such as IR42, MR71, MR73, MR77 and MR84 were widely recommended and adopted.

106

- **One-month fallow period** A one-month fallow period was implemented in the MUDA Scheme in 1984 to interrupt the otherwise continuous cropping. This break was effective in reducing the vector population.

- **Synchronized planting** Every effort was made to ensure that farmers planted synchronously so as to minimize vector and virus build-up.

- **Destruction of tungro sources** Farmers were urged to burn ratoons, stubbles and volunteer seedlings to reduce survival of hosts for GLH and tungro. Dry-ploughing after harvest, roguing and herbicide spraying were also recommended.

- **Chemical control** Use of insecticides was recommended for vector control. In seedling nurseries, control of vectors was sought by incorporating carbofuran (1 kg a.i./ha) into nursery beds or by spraying with insecticides such as BPMC (2-sec-butylphenyl methylcarbamate), carbaryl or MIPC (Isoprocarb). After transplanting vectors were controlled by spraying with these insecticides. Farmers were advised to use insecticides judiciously, using timely applications and spot-spraying any localized groups of infected plants. Currently, if spraying is justified, buprofezin and isoprocarb or ethofenprox are recommenced so as to maintain natural enemies.

- **Other cultural control** It was recommended that nurseries should be located away from light sources as GLH is attracted to light. Infected seedlings should be rogued and destroyed as quickly as possible. Movement of seedlings from infected areas was strongly discouraged. Farmers were also advised to control the weeds along the bunds and to avoid excessive nitrogen application. Moreover, the value of natural enemies in reducing GLH populations was also considered and indiscriminate use of insecticide was discouraged so as to conserve them.

DISCUSSION

The management of tungro in Malaysia has been implemented successfully following an integrated pest management (IPM) approach. Close co-operation between the various government agencies and the active participation of the farmers have ensured the success of the campaign. Although tungro is currently well under control, every effort is being made to maintain this situation. An expert system for the control of tungro has been designed by MARDI to assist the extension staff to make decisions on whether control measures are needed (Figure 4). Future IPM programmes for tungro in Malaysia will focus on the refinement and improvement of the existing programme. The increasing awareness of the need for environmental conservation and sustainable agriculture will be the main motivating factor in the development of IPM for tungro.

REFERENCES

DEPARTMENT OF AGRICULTURE (1983) Kempen kawalan penyakit merah di Semenanjung Malaysia. Laporan Suku Tahun, Cawangan Pemeliharaan Tanaman, Kuala Lumpur.

HIBINO, H., ROECHAN, M. and SUDARISMAN, S. (1978) Association of two types of virus particles with penyakit habang (tungro disease) of rice in Indonesia. *Phytopathology*, **68**: 1412–1416.

HIBINO, H., SALEH, N. and ROECHAN, M. (1979) Transmission of two kinds of rice tungro-associated viruses by insect vectors. *Phytopathology*, **69**: 1266–1268.

HIBINO, H., ISHIKAWA, K., OMURA, T., CABAUATAN, P.Q. and KOGANEZAWA, H. (1991) Characterization of rice tungro bacilliform and rice tungro spherical viruses. *Phytopathology*, **81**: 1130–1132.

ISHIHARA, T. and KAWASE, E. (1968) Two new Malayan species of the genus *Nephotettix* (Hemiptera: Cicadellidae). *Applied Entomology/Zoology*, **3**: 119–123.

LOCKARD, R.G. (1959) *Mineral Nutrition of the Rice Plant in Malaya, with special reference to* Penyakit Merah Bulletin No. 198. Kuala Lumpur: Department of Agriculture.

LIM, G.S. (1972) Studies on *penyakit merah* disease of rice, 3. Factors contributing to an epidemic in North Krian, Malaysia. *Malaysian Agricultural Journal*, **48**: 278–294.

LING, K.C. (1966) Non-persistence of the tungro virus in its leafhopper vector *(Nephotettix impicticeps). Phytopathology,* **56**: 1252–1256.

OMURA, T., SAITO, Y., USUGI, T. and HIBINO, H. (1983) Purification and serology of rice tungro spherical virus and rice tungro bacilliform virus. *Annals of Phytopathological Society of Japan,* **49**: 73–76.

OU, S.H. (1965). Rice diseases of obscure nature in Tropical Asia with special reference to *Mentek* disease in Indonesia. *International Rice Commission Newsletter,* **14**: 4–10.

OU, S.H. (1985) *Rice Diseases.* 2nd edn. Kew, UK: Commonwealth Mycological Institute.

OU, S.H. and GOH, K.G. (1966) Further experiments on *penyakit merah* disease in Malaysia. *International Rice Commision Newsletter,* **15**: 31–32.

PERANGKAAN PADI (1982) Malaysian Ministry of Agriculture.

SAITO, Y. (1977) Interrelationship among waika disease, tungro and other similar diseases of rice in Asia. *Tropical Agricultural Research Centre,* **10**: 129–135.

SINGH, K.G. (1969) Virus–vector relationship: penyakit merah of rice. *Annals of the Phytopathological Society of Japan,* **35**: 322–324.

SINGH, K.G. (1971) Recent progress in rice insect research in Malaysia. *Proceedings of the Symposium on Rice Insects, 1971.* Tropical Agricultural Research Centre, Ministry of Agriculture and Forestry, Japan.